The History of Oil

Tracing the Black Gold

CRAFTED BY SKRIUWER

Dear Reader,

Imagine standing by a natural oil seep in ancient Mesopotamia, watching builders mix bubbling bitumen to waterproof the great ziggurats. Or journeying with early Chinese drillers as they pierce the earth with bamboo rigs, seeking the "burning water" that will light their lamps. That's where this book begins: not with technical charts or dry data, but with the human ingenuity and drama behind oil's timeless story.

As you turn these pages, you'll walk alongside:

- *Ancient engineers using bitumen to fortify the walls of Babylon and preserve Egyptian mummies for the afterlife (Chapters 1–2)*
- *Alchemists in the Islamic Golden Age, unlocking the secrets of distillation and lighting the lamps of Baghdad (Chapter 5)*
- *Whalers and adventurers of the 18th century, stumbling upon "tar springs" in the Americas—and sensing their potential (Chapter 11)*

We've structured this journey to show how yesterday's breakthroughs shape today's world:
🔥 *How ancient sealing techniques foreshadowed modern petroleum engineering*
🔥 *How the alchemical stills of medieval Persia evolved into today's refineries*
🔥 *How early environmental concerns in the 1800s echo in our climate debates*

This isn't just history – it's the backstory of modern civilization. When you read about:

- *The role of oil in conquest and trade (Chapters 4, 7, 8), you'll see the roots of today's geopolitical struggles*
- *The rise of refining and lubrication (Chapters 14–15), you'll understand the making of our industrial age*
- *Early regulations and community protests (Chapters 17–18), you'll recognize the origins of today's energy ethics*

Why this format?

- *Clear storytelling turns complex innovation into gripping narrative*
- *Global perspectives honor both Eastern and Western advancements*
- *Thematic chapters help you trace the unbroken thread from past to present (see full contents inside)*

So grab coffee (or something stronger). Let's journey from bitumen pits to oil derricks—through discovery, risk, revolution, and consequence. May these stories not only inform you... but transform how you see the world that oil built.

With anticipation,
Skriuwer

At **Skriuwer**, we're more than just a team—we're a global community of people who love books. In Frisian, "Skriuwer" means "writer," and that's at the heart of what we do: creating and sharing books with readers worldwide. Wherever you are in the world, **Skriuwer** is here to inspire learning.

Frisian is one of the oldest languages in Europe, closely related to English and Dutch, and is spoken by about **500,000 people** in the province of **Friesland** (Fryslân), located in the northern Netherlands. It's the second official language of the Netherlands, but like many minority languages, Frisian faces the challenge of survival in a modern, globalized world.

We're using the money we earn to promote the Frisian language.

For more information, contact : **kontakt@skriuwer.com** (www.skriuwer.com)

Historical Research Methodology

This work synthesizes information from:

- **Primary Sources**

  - Government archives and historical records
  - Period correspondence and eyewitness accounts
  - Contemporary newspapers and publications

Disclaimer:

The images in this book are creative reinterpretations of historical scenes. While every effort was made to accurately capture the essence of the periods depicted, some illustrations may include artistic embellishments or approximations. They are intended to evoke the atmosphere and spirit of the times rather than serve as precise historical records.

"History is the version of past events that people have decided to agree upon."

– Napoleon Bonaparte

THE MOST EXPLOSIVE QUOTES ABOUT OIL HISTORY

'This is the stuff that will make the world's lamps burn brighter.'

— George Bissell (1855, upon seeing oil samples)

'Stop pumping? The oil is there to be taken.'

— John D. Rockefeller (c. 1880s, on production concerns)

'I should have wished to be an oil man.'

— Winston Churchill (1911, as First Lord of the Admiralty)

'The oil can is mightier than the sword.'

— Everett Lee DeGolyer (c. 1940s, prominent geologist)

'What's good for Standard Oil is good for the nation.'

— Frank Abrams (1950s, Chairman of Standard Oil)

'The law of capture is the fundamental rule of oil and gas.'

— Legal Doctrine (Early 20th Century)

'We will drown you in oil.'

— Mikhail Gorbachev (1986, to West German officials)

'We will pump more oil to bring down the price.'

— Sheikh Ahmed Zaki Yamani (1970s, Saudi Oil Minister)

OIL HISTORY AT A GLANCE: THE POWER BENEATH

A Visual Timeline of the History of Oil

Antiquity – 1400s: Ancient Foundations

- *Earliest use of bitumen in Mesopotamian construction (3000 BCE)*
- *Chinese drill first oil wells using bamboo (347 CE)*
- *Oil used in Greek fire, mummification, and medicine*

1500s–1700s: Global Exploration & Early Interest

- *Oil seepages documented in the Americas by explorers*
- *Advancements in distillation by Islamic and European alchemists*
- *First oil cartels: "The Fellowship of Lighters" controls Caspian sea oil (1594)*

1800–1849: Proto-Industry & Innovation

- *First commercial oil well dug in Burma (hand-dug, 1798)*
- *New scientific curiosity in geology and "rock oil"*
- *Kerosene refined from coal, creating demand for lamp fuel*

1850–1899: The Dawn of the Modern Era

- *Drake's first modern oil well in Titusville, Pennsylvania (1859)*
- *John D. Rockefeller incorporates Standard Oil (1870)*
- *Internal combustion engine patented (1886)*
- *Spindletop gusher sparks Texas oil boom (1901)*

1900–1945: The Age of Fuel & Global Power

- *Ford Model T introduces mass automobile ownership (1908)*
- *WWI mechanizes warfare, making oil a strategic military asset*
- *OPEC precursor formed as oil-rich nations assert control (1928 Red Line Agreement)*

1945–1973: The Postwar Oil Order

- *Mass expansion of suburbs and car culture*
- *Rise of "Seven Sisters" oil conglomerates*
- *U.S. becomes a net oil importer (1948)*
- *Oil surpasses coal as world's primary energy source*

1973–1990: Shocks & Sovereignty

- *OPEC oil embargo triggers global energy crisis (1973)*
- *Prudhoe Bay, Alaska discovery (largest in US history) (1968)*
- *Exxon Valdez spill highlights environmental cost (1989)*

1990–Present: Globalization & Uncertainty

- *Peak of conventional oil production debated (Peak Oil theory)*
- *Deepwater Horizon disaster (2010)*
- *Shale revolution fracking makes US top global producer (2010s)*
- *Global push for energy transition away from fossil fuels*

"We will squeeze the lemon until the pips squeak."

— Calouste Gulbenkian (c. 1920s, on negotiating oil concessions)

BEFORE WE START, HERE ARE 50 CRAZY, INTERESTING AND DISTURBING FACTS ABOUT HISTORY OF OIL

💉 In the 1850s, "snake oil" salesmen were a real thing. They sold petroleum as a fraudulent miracle cure, decades before it was used as fuel.

⚰️ Ancient Egyptians used oil not for energy, but for eternity—bitumen was a key ingredient in their mummification process, preserving the dead for thousands of years.

💰 The world's first billionaire, John D. Rockefeller, became one by controlling 90% of the American oil industry, using ruthless tactics that would be illegal today.

🔥 "Greek Fire," a terrifying ancient napalm used by the Byzantine Empire, was so effective its exact recipe—believed to contain oil—was a state secret lost to history.

💉 The first modern oil well was drilled in 1859 because we needed light; whale oil was becoming too expensive and scarce from overhunting.

💣 During WWII, a secret Allied plan proposed flooding Germany with fake, explosive coal briquettes made of oil and coal dust to destroy Nazi boilers and factories.

🚂 The first oil tank cars were just modified wooden wine barrels strapped to railcars, often leaking and causing spectacular, deadly fires on the tracks.

🗺️ In 1928, the heads of the world's top oil companies literally drew a "Red Line" on a map of the Middle East to carve up oil rights amongst themselves.

🦟 The pesticide DDT, famously condemned in Silent Spring, is derived from petroleum, linking the oil industry to one of history's greatest environmental controversies.

🤝 Standard Oil secretly owned the companies that were supposedly its biggest competitors, creating an illusion of choice while maintaining a total monopoly.

💡 Kerosene was initially a cheap byproduct of refining oil for lamp fuel; gasoline was considered a useless and dangerous waste product often dumped into rivers.

🛢️ The first oil boom in the Americas wasn't in Texas—it was in Ontario, Canada, in 1858, a full year before Drake's well in Pennsylvania.

🧪 The development of plastics was a direct result of the oil industry's desperate search for new uses for its endless supply of petroleum byproducts.

🚗 Henry Ford's first car, the Quadricycle, ran on ethanol. He only pivoted to gasoline because John D. Rockefeller's kerosene was so cheap and abundant.

🕵️‍♂️ The CIA's first-ever covert operation, codenamed "Ajax," wasn't about spies; it was about oil—orchestrating a coup in Iran to regain control of British oil assets in 1953.

Vaseline, a pure petroleum jelly, was sold in the 1870s with the outrageous claim that ingesting a spoonful a day would protect your digestive tract.

An oil gusher in California in 1910 was so massive it created a river of crude that killed men, drowned animals, and spilled more oil than the Exxon Valdez.

Lead was deliberately added to gasoline to stop engine knocking, despite known fatal health effects, making generations breathe neurotoxic leaded fumes for decades.

In the late 1800s, oil wildcatters used "doodlebugs"—fraudulent divining rods—to convince investors they could magically find oil underground.

The term "hot oil" in the 1930s didn't mean stolen crude; it meant oil produced in excess of legal quotas, sparking a violent "Oil War" in Texas.

The first offshore oil rig wasn't built in the Gulf of Mexico; it was built off the coast of California in 1896, using a wooden pier extended into the Pacific.

The 1973 OPEC oil embargo was so effective it caused a stock market crash and led to the invention of the modern financial "futures" market for oil.

Soap is deeply tied to oil; the industrial process for making modern soap bars from animal fat was revolutionized using petroleum-derived chemicals.

The vibrant synthetic colors in modern art were born from the oil industry, which provided the chemical base for new pigments previously impossible to create.

During the Civil War, Confederate soldiers would raid Pennsylvania oil fields not for fuel, but to use crude oil as a primitive medicine for their wounds.

The ancient city of Carthage was famously razed and salted by the Romans, but its walls were first waterproofed with bitumen—a product of oil.

The founder of modern geophysics was a convicted murderer; Albert Einstein's son later worked for him, helping the oil industry find reserves using sound waves.

Nylon, the first completely synthetic fiber, was created from coal, air, and water by DuPont, marking the shift towards a petrochemical-based world.

The iconic "oil derrick" is named after a type of gallows, because the wooden structure resembled the device used to hang a famous 17th-century English executioner.

The aspirin you take today is made from benzene, a hydrocarbon derived from petroleum, not from the willow bark used in ancient times.

The whaling industry was completely obliterated almost overnight by the oil industry, as cheap kerosene replaced expensive whale oil in lamps.

🧛‍♂️ The original "vampire" myths in Europe may be linked to oil; peasants opening graves sometimes found bodies preserved in crude oil seepage, mistaken for the undead.

💉 The first plastic, Bakelite, was made from phenol, a compound derived from coal tar and later petroleum, launching the Age of Plastic.

🗣 The famous slogan "See the USA in your Chevrolet" was a direct response to the 1956 Suez Crisis, which threatened oil supplies and promoted domestic travel.

🦠 Oil-eating bacteria exist naturally in the ocean. The Deepwater Horizon spill showed they can bloom to consume a spill, but their waste creates ocean dead zones.

🧴 The world's first major oil spill happened not from a tanker, but from a gusher in California that spilled 9 million gallons over 18 months before it was capped.

🎪 The first cars were considered a nuisance because they scared horses; their adoption only exploded after oil became cheap enough to make them practical.

🧠 The scientist who discovered the neurotoxicity of lead in gasoline was hired by General Motors to prove its safety; when he found the opposite, they fired him and hid his work.

🛢 Saddam Hussein set over 600 Kuwaiti oil wells on fire as a scorched-earth tactic in 1991, creating apocalyptic fires that took 9 months to extinguish.

🧪 Synthetic rubber, vital for WWII tires and equipment, was invented because the Japanese military cut off the Allies' access to natural rubber plantations in Asia.

🚢 The switch from coal to oil-powered British warships before WWI was so secretive that Germany couldn't understand how the UK's ships suddenly became so much faster.

🌎 The famous "Spindletop" gusher was found not by a geologist, but by a self-taught miner who was repeatedly told by experts that he was drilling in the wrong place.

🍔 Synthetic fertilizers that feed half the world's population are made from natural gas, a hydrocarbon often found and produced alongside oil.

🕵️‍♀️ During the Cold War, geologists' maps of oil potential were considered top-secret documents, as valuable to spies as missile schematics.

🦟 The pesticide that enabled the agricultural "Green Revolution" and saved millions from malaria was made from petroleum, creating a paradoxical dependency.

🧪 The distinctive smell of "new car" is not leather or fabric; it's the off-gassing of dozens of petroleum-based plastics, adhesives, and solvents in the interior.

🛢️ The first oil pipeline, built in 1865, was so controversial that teamsters who hauled oil in wagons would shoot holes in it to put it out of business.

👨‍🚀♂️ The suit worn by the first man on the moon, Neil Armstrong's spacesuit, was made of synthetic materials like neoprene and nylon—all petroleum-based.

🧪 The entire field of organic chemistry, which studies carbon-based life, was built on the back of experimenting with the hydrocarbons found in coal tar and oil.

🌍 The Great Acceleration—the dramatic spike in human population and consumption since 1950—is plotted on a graph that perfectly mirrors the curve of global oil production.

TABLE OF CONTENTS

CHAPTER 6: OIL ACROSS MEDIEVAL EUROPE

- *Limited but notable uses of bitumen*
- *Transmission of oil knowledge from the Middle East*
- *Role of monasteries in preserving technical texts*
- *Growth of small-scale craft industries*

CHAPTER 7: EARLY TRADE ROUTES AND THE RENAISSANCE INFLUENCE

- *Expansion of global exploration*
- *Renewed scientific curiosity in natural resources*
- *Bitumen in European shipbuilding and repairs*
- *Stimulus for further experimentation*

CHAPTER 8: OIL IN THE AGE OF EXPLORATION

- *Discovery of new seep sites in the Americas*
- *Competition between European powers*
- *Early documentation of oil's potential*
- *Impact of maritime journeys and resource claims*

CHAPTER 9: PROTO-INDUSTRIAL USES OF OIL BEFORE 1800

- *Emerging industrial crafts and rudimentary machinery*
- *Use of oil for lubrication and minimal-scale lighting*
- *Shift from local to broader economic interest*
- *Precursors to modern drilling ideas*

CHAPTER 10: GROWING SCIENTIFIC INTEREST IN GEOLOGY AND OIL

- *Early geological surveys and theories*
- *Mapping likely regions for oil seepage*
- *New scientific instruments for resource exploration*
- *Influence on developing scientific communities*

CHAPTER 11: ADVENTURERS AND EXPLORERS SEEKING OIL (17TH–18TH CENTURIES)

- *Personal accounts of black, tar-like springs*
- *Risk and reward of early oil ventures*
- *Cultural fascination with "miracle substances"*
- *Shift from folk knowledge to systematic search*

CHAPTER 12: OIL IN THE LIVES OF EARLY COLONIAL SOCIETIES

- *Interactions with indigenous knowledge of oil*
- *Use of oil for lamps, medicine, and heating*
- *Challenges in gathering enough oil for communities*
- *Role of local governance in regulating resources*

CHAPTER 13: OIL EXTRACTION TECHNIQUES IN THE EARLY 19TH CENTURY

- *Shallow pit digging and crude well technology*
- *Mechanical innovations to increase output*
- *Early commercial selling of raw oil*
- *Case studies of pioneering attempts*

CHAPTER 14: THE RISE OF EARLY OIL REFINING (19TH CENTURY)

- *Basic processes for separating oil fractions*
- *Small-scale operations and local refineries*
- *Market demands for lamp fuel and lubricants*
- *Gradual improvement in refining efficiency*

CHAPTER 15: OIL AND TRANSPORTATION SHIFTS OF THE 19TH CENTURY

- *Experiments with oil-based engines and steam power*
- *Development of lubricants for railroads and ships*
- *Transition from traditional fuels to emerging oil usage*
- *Practical challenges in distribution and storage*

CHAPTER 16: OIL AND POLITICAL POWER BEFORE THE LATE 19TH CENTURY

- *Resource control by local leaders and small states*
- *Early negotiations and trade agreements*
- *Economic influence on regional conflicts*
- *Social hierarchy and ownership of oil lands*

CHAPTER 17: CULTURAL AND SOCIAL IMPACT OF OIL IN THE 19TH CENTURY

- *Public perception of "rock oil" and innovation*
- *Growth of local industries and job creation*
- *Evolving attitudes toward environmental effects*
- *Early debates and protests over oil expansion*

CHAPTER 18: ENVIRONMENTAL VIEWS AND EARLY CONCERNS

- *Noticing the effects on water and soil*
- *First attempts at regulating seepage sites*
- *Health impacts on local communities*
- *Emerging calls for responsible usage*

CHAPTER 19: GLOBAL SEARCHES FOR OIL IN THE LATE 19TH CENTURY

- *Increased curiosity beyond known oil fields*
- *Entrepreneurial ventures and speculation*
- *International communication of new discoveries*
- *Laying groundwork for future major booms*

CHAPTER 20: THE THRESHOLD OF THE MODERN OIL ERA

- *Early successes leading to more ambitious drilling*
- *Refining breakthroughs hinting at large-scale production*
- *Societal changes driven by new energy options*
- *Transition to the industrial age and beyond*

CHAPTER 1: ANCIENT BEGINNINGS OF OIL USE

Introduction

The history of oil did not start in the modern age. It began many thousands of years ago, long before people even imagined the world we live in today. Ancient people came across oil in its natural state when it seeped through the ground. They might have seen thick, black, sticky pools on the surface or a shiny film on water in certain areas. Curious individuals tested it, discovered it could burn, and found that it had many other special properties. This discovery would shape how they built their houses, protected their boats, and lit up dark spaces.

In ancient times, people did not have the scientific knowledge or the technical tools we have today. They did not know that oil formed from the remains of ancient plants and tiny organisms over millions of years. They did not use the word "petroleum" or talk about hydrocarbons. All they knew was that this sticky substance could help them in their daily lives. Over time, they learned ways to gather it from the ground or from natural seeps. Through trial and error, they found simpler methods to extract or refine it for their needs.

In this chapter, we look at how oil was first discovered and utilized by the earliest human civilizations. We explore how it caught their attention, the ways they used it, and how these ancient uses laid the foundation for later development. We do not dive too far ahead in history. Instead, we focus on the earliest clues we have from archaeology and written records. These glimpses show us that oil had already played a role in some of humanity's oldest achievements.

Oil's First Encounters

Oil, in the form of natural seepage, appeared in many regions of the world. In certain lands, like what is now the Middle East, you could find areas where the ground had cracks or fissures that leaked a black, tar-like substance. In other places, such as present-day California, ancient peoples also discovered tar pits. Animals sometimes got trapped in these pits, leaving behind fossils that modern researchers use to learn about the past. But for ancient people, these pits were both a hazard and a source of useful material.

The earliest direct evidence of human interaction with oil may be the use of bitumen, a thick form of crude oil, for building and waterproofing. Archaeologists have found that bitumen was used to seal boats and buildings in many parts of the ancient world. This practice might have started as soon as people realized that applying a layer of this sticky substance kept out water and made their structures more stable. Over time, knowledge of how to gather bitumen spread from one group to another.

Because written records from the truly ancient periods (like the late Stone Age) are sparse, we rely on artifacts, remains of settlements, and geological clues. For example, if archaeologists find lumps of bitumen attached to pottery shards in a prehistoric site, it suggests that the people living there understood how to collect and store that substance. Some lumps show signs of being melted or heated, which indicates a controlled use of fire to soften or refine the bitumen.

Uses in Daily Life

Early societies were practical. If something was useful, they found a way to incorporate it into their daily routines. Oil and bitumen had many possible purposes:

1. **Waterproofing**: One of the simplest and most common uses was sealing boats, rooftops, and containers. If you coat a surface in bitumen, water cannot easily penetrate it. This was vital for keeping grain dry, especially during rainy seasons or in flood-prone regions.
2. **Adhesive**: Oil, in a thicker or partially dried form, could be sticky enough to function as glue. Ancient craftsmen might have used it to bond materials, seal containers, or attach decorative items to surfaces.

3. **Lighting**: When some form of crude oil or pitch was set on fire, it produced light. While early lamps more commonly used animal fat or plant oils, it is possible that some communities tested crude oil or bitumen lumps to create small flames.
4. **Symbolic or Religious Purposes**: Substances that burned brightly or had distinctive smells often found their way into rituals. We lack detailed records for many ancient cultures, but it is not a stretch to think that burning black tar in ceremonies might have held symbolic significance.
5. **Medicinal and Preservative Uses**: Certain ancient societies used bitumen or crude oil-like substances on wounds, believing they had healing properties. These substances might also have helped preserve wood or other materials, slowing decay.

Regional Variations

Different parts of the ancient world developed their own methods and cultural views regarding oil. For example:

- **Middle East**: Rich in natural seep sites, areas near the Tigris and Euphrates Rivers had many bitumen pits. Early peoples here likely used bitumen for constructing city walls and for religious structures.
- **Central Asia**: There are indications that tribes in regions like present-day Turkmenistan found sticky substances that came from the ground. They may have used them for simple tasks like caulking wooden items.
- **North America**: Native American tribes who lived near tar pits or "oil springs" used thick oil for waterproofing canoes or as a skin ointment. Oral traditions in certain tribes mention the unusual substances coming from the earth.

Ancient societies were highly adaptive. Even if they did not fully understand the composition of oil, they recognized its potential. In many cases, the use of oil existed alongside other natural resources, such as clay, stone, or reeds.

Evidence from Archaeology

Much of what we know about early oil use is pieced together from artifacts. Archaeologists have discovered:

- **Pottery Residues**: Pottery fragments with residues that test positive for bitumen.
- **Sealed Structures**: The remains of ancient buildings with layers of black, tar-like substance in the foundation or on the walls.
- **Tools and Containers**: Tools sometimes had a black adhesive or protective coat. Containers used for transporting goods could be lined with bitumen to prevent leaks.

From these clues, we can guess the range of oil's early applications. Sometimes, the presence of bitumen in a site far from natural seep sources shows that a trade network existed. People transported lumps of bitumen or traded for it, meaning it had enough value to move from one place to another. Over time, small communities that had easy access to seep sites might have specialized in collecting and selling or bartering this resource.

Early Myths and Legends

In addition to practical uses, oil or tar may have appeared in myths and folklore. Ancient peoples often told stories about the natural world to explain its wonders and dangers. A place where the earth seemed to sweat black goo might become the focus of a legend about gods or spirits who lived underground. While we lack exact details for many of these stories, it is reasonable to believe that oil seepage sites became local landmarks around which tales were spun.

Some of the oldest stories from Mesopotamia mention the importance of certain substances in building and protection. Though not always directly stating "oil," references to pitch or bitumen appear in cuneiform texts. These texts might describe how a king sealed his boat or how an offering was burned in a sacred fire. Over time, these references helped shape cultural attitudes, turning oil-based substances from a mere curiosity into an accepted part of religious and civic life.

Limitations of Early Methods

While ancient peoples learned a lot about oil, their methods for collecting and refining it were rudimentary compared to later developments. Gathering oil might have involved scooping up bits of tar from shallow pools or waiting for seepages to accumulate. There was little concept of drilling deep into the earth with advanced tools. Furthermore, because the scale of use was small, there was no immediate drive to discover more efficient methods.

Refining, if we can call it that, involved heating or boiling the tar to separate some of the lighter components or to soften it for easier application. The fact that bitumen ignites easily meant that ancient refiners had to be cautious. Fires might have been frequent, which taught them respect for the substance's flammable nature. Any deeper analysis of oil's composition was impossible given the era's limited scientific knowledge.

Cultural Exchange

Trade routes that connected distant lands played a huge role in spreading knowledge about oil. People in one region might learn from travelers or merchants that in a faraway place, black tar was used for a particular construction technique. They might then try to replicate that technique at home. Over many generations, this process of sharing knowledge led to a gradual but widespread awareness of oil's potential.

For example, caravans traveling across the Middle East carried not only goods like spices and textiles but also stories and ideas. A merchant from a region near the Dead Sea might talk about the thick asphalt or bitumen found floating on the water's surface. Another merchant from the Persian Gulf might describe natural "fire wells" where gas escaped the ground and lit up the night. This exchange of information set the stage for further experiments in how to use oil effectively.

The Significance of Ancient Oil Use

It is tempting to think of oil's significance as purely modern. But even in ancient times, oil-based substances contributed to major events and everyday survival. By letting humans waterproof their boats, it helped them travel rivers and seas more confidently. By sealing roofs, it protected them from the elements. By lighting small lamps, it extended the time they could be active after sunset. On a more subtle level, it shaped myths, influenced local economies, and possibly even determined political decisions for tribes that controlled seepage sites.

Ancient uses of oil also served as a stepping stone. They showed that humans were ready to utilize the resources at their disposal in clever ways. As the centuries passed, the knowledge of how to gather and handle oil did not vanish. Instead, it evolved, picking up new techniques and applications from each culture that engaged with it. While no single ancient society fully unlocked oil's transformative power, they all contributed pieces of the puzzle that future generations would assemble.

CHAPTER 2: OIL IN EARLY CIVILIZATIONS OF MESOPOTAMIA AND EGYPT

Introduction

Mesopotamia and Egypt rank among the earliest and most influential civilizations in history. Located in regions where natural seepages of bitumen and crude oil were not uncommon, these societies made advanced use of oil-based substances. The Tigris-Euphrates River Valley (often called the "Cradle of Civilization") had significant bitumen deposits, while ancient Egypt, centered around the Nile, discovered ways to incorporate bitumen into religious and daily life. Understanding their innovative uses of oil shows us how valuable this resource had become in the ancient world.

This chapter looks at how these two great civilizations encountered oil, how they organized its extraction, what they used it for, and how it featured in their trade networks. We see that oil was not only a tool for construction but also appeared in spiritual rituals and cultural customs. By examining these societies' handling of bitumen and crude oil, we get a clearer picture of oil's importance in shaping large-scale projects like pyramids, massive temples, and city fortifications.

Mesopotamia: The Cradle of Civilization and Oil

Mesopotamia, the region between the Tigris and Euphrates rivers (modern-day Iraq and parts of Syria, Turkey, and Iran), had fertile land that supported some of the earliest known cities. Civilizations like the Sumerians, Akkadians, Babylonians, and Assyrians rose and fell in this area over centuries. Given the presence of natural bitumen seepages near the rivers, it is no surprise that Mesopotamian builders made extensive use of it.

Construction and Architecture

One of the most striking aspects of Mesopotamian civilization was the building of massive structures using mud bricks. These bricks were sturdy but vulnerable to water damage, especially during floods. To protect walls and foundations, Mesopotamian engineers discovered that coating bricks with bitumen could repel water. Evidence shows that the lower sections of city walls were sometimes lined with bitumen for added durability.

Ziggurats—large temple towers built in step-like shapes—often had layers of bitumen in their construction. Although the core of the structure was mud brick, certain layers between bricks or around the base included bitumen to keep moisture from eroding the building's foundation. This technique demonstrates a refined understanding of how to preserve large-scale structures in a climate prone to seasonal flooding.

Water Management and Canals

Mesopotamians were known for their complex irrigation systems. They dug canals to direct water from the rivers to their fields. Here, too, bitumen found an application. By lining canals with bitumen, they could reduce seepage and control water flow more efficiently. This was crucial for supporting agriculture, which in turn sustained the population and spurred the growth of cities.

As society expanded, the demand for food increased. The ability to manage water effectively was a mark of a sophisticated civilization. Bitumen helped them in this endeavor by reducing water loss. Without a secure supply of water, fields would dry out, and the city populations would face famine. So in many ways, bitumen supported the very growth of urban life in Mesopotamia.

Religious and Ceremonial Uses

Mesopotamians were deeply religious, with temples and rituals forming the center of daily life. Bitumen or crude oil might have been used in certain offerings or burned in lamps during ceremonies. Cuneiform texts sometimes refer to the use of pitch or tar in sacred contexts. While details can be scarce, references in ancient tablets suggest that priests or temple attendants would use oil-based substances as part of purification rituals or temple upkeep.

Furthermore, certain texts mention the application of bitumen-like substances to statues of gods. This might have been to protect them from weather or to anoint them in religious ceremonies. The thick, black color could have had symbolic meanings tied to fertility, the underworld, or protective powers.

Trade and Economy

Trade was vital in Mesopotamia. Cities like Ur, Uruk, Babylon, and Nineveh became centers of commerce, connecting with distant regions. Bitumen itself emerged as a trade good, especially in areas that lacked local seepages. Mesopotamia's central location meant it could acquire metals, precious stones, and other resources from abroad and exchange them for the bitumen produced near the rivers.

Merchants would transport lumps of solidified bitumen in baskets or skins. Because it was a valuable resource for construction, waterproofing, and possibly medicinal uses, demand remained steady. The wide reach of Mesopotamian trade networks ensured that knowledge of bitumen's utility spread, influencing neighboring civilizations.

Cultural Attitudes

The Mesopotamians did not view bitumen as just another building material. It had a special place in their worldview, evidenced by how often it appears in myths and epic stories. In the Epic of Gilgamesh, for instance, references to using tar to seal a boat during a great flood mirror the story's emphasis on survival and rebirth. That detail shows that bitumen was recognized as essential for saving life in a deluge.

With time, the repeated use of bitumen in iconic structures, religious rituals, and practical needs shaped how Mesopotamians perceived it. It was not merely a commodity; it symbolized innovation, protection, and the ability to thrive in a challenging environment.

Egypt: The Gift of the Nile and the Use of Bitumen

Meanwhile, in ancient Egypt, society flourished along the Nile River. Known for their monumental pyramids, grand temples, and remarkable mummification practices, the Egyptians also encountered natural seepages of oil in parts of the region, though less frequently than in Mesopotamia. Still, they learned to transport and utilize bitumen from nearby areas, such as the Dead Sea region (sometimes called Lake Asphaltites in ancient texts).

Mummification and Ritual

One of the most famous uses of bitumen in Egypt is in the process of mummification. The word "mummy" itself is believed to come from the Persian or Arabic word for bitumen ("mumiya"), pointing to the blackish substance used in preserving bodies. While the classic mummification process involved removing organs, drying the body with natron, and wrapping it in linen, bitumen was sometimes incorporated as part of the embalming and sealing process.

Bitumen gave mummies a dark color, and its preservative qualities helped slow decomposition. Over centuries, the Egyptians refined their mummification techniques, making the bodies more resilient against decay. This practice was deeply tied to their beliefs about the afterlife, making bitumen a substance of great spiritual importance.

Construction and Sealing

Although Egyptian architecture often featured stone, bitumen still had its place. In some temple complexes and administrative buildings, bitumen served as a sealant to protect foundations from water damage, especially near the Nile's floodplain. Boats traveling the Nile might also have used bitumen to seal wooden hulls. Given the essential role of the river in Egyptian life—both for transportation and agriculture—waterproofing was critical.

Some archaeological sites show evidence of bitumen used alongside mortar. While less common than in Mesopotamia, these findings confirm that Egyptians also recognized how bitumen could shield structures from moisture. In a land where the Nile's annual flood was a regular occurrence, preventing water intrusion was important for preserving valuable buildings and goods.

Trade Connections with the Near East

Egypt relied on trade routes for many goods, including wood, metals, and spices not available within its borders. Bitumen was one of these traded items. If local seepages were not enough, Egyptians imported bitumen from regions around the Red Sea or further northeast. This cross-regional commerce fostered cultural exchange, and Egyptians likely learned new methods or uses for bitumen from traders who traveled across the Middle East.

Documents from the New Kingdom period describe shipments of tar-like substances coming into Egypt. These might have been used for temple rituals or for embalming. The link between bitumen trade and religious practices was strong because of its essential role in ensuring the safe passage of pharaohs and high-status individuals into the afterlife.

Symbolic Meaning

Egyptian religion was filled with symbolism. Many substances had magical or protective aspects. Bitumen, with its dark hue and preservative properties, could represent regeneration or the power to ward off decay. The land of Egypt was often seen as a place where life prevailed against the chaotic forces of the desert. Substances that could preserve life (even in death) naturally took on a mystical aura.

In some texts, the black color was associated with fertility, linked to the rich black silt deposited by the Nile floods. Bitumen's darkness might have tied it symbolically to the fertile soil that sustained Egypt's agricultural wealth. Thus, using bitumen in burials could have connected the deceased to the regenerative power of the land, reinforcing beliefs about rebirth in the afterlife.

Overlapping Themes in Mesopotamia and Egypt

While distinct, Mesopotamian and Egyptian uses of oil-based substances share notable similarities:

1. **Practical Building and Waterproofing**: Both civilizations applied bitumen to protect structures from water damage.
2. **Religious Significance**: Bitumen or crude oil sometimes played roles in rituals or ceremonies, highlighting its perceived power.
3. **Trade and Economic Value**: Regions without natural seepages imported bitumen from areas that had it.
4. **Preservation and Protection**: Whether preserving mummies or safeguarding buildings, oil-based substances served as a protective shield against decay or damage.

These commonalities show that once people recognized oil's benefits, the knowledge spread and adapted to different cultural settings. Both civilizations left behind grand monuments, texts, and artifacts that give us a window into how they viewed and utilized oil in everyday and sacred contexts.

Interaction and Exchange

Mesopotamia and Egypt did not exist in total isolation from one another. There were periods of trade, diplomacy, and even warfare between them. When goods and ideas traveled along these routes, so did methods for handling oil. Mesopotamian craftspeople might have influenced Egyptian practices, and vice versa. Over the centuries, these interactions ensured that bitumen, once a local resource, became something recognized across the Near East for its usefulness and spiritual connotations.

Egypt's interest in bitumen from places near the Dead Sea (often referred to as a major source of asphalt in the ancient world) is an example of cross-cultural trade. Mesopotamians who lived closer to that region might have had direct control or knowledge of bitumen extraction sites. This created opportunities for merchants who profited by transporting lumps of tar along caravans or up the Nile.

Innovations and Techniques

Both civilizations developed ways to make bitumen easier to work with. They might have heated it or mixed it with other substances like straw or sand to change its consistency. Ancient texts sometimes refer to different grades of bitumen, possibly indicating some level of refinement. By adjusting the heat or the mixture, craftspeople could create a product that was harder or softer, depending on the need.

While we do not have exact recipes for these mixtures, archaeological evidence of bitumen lumps containing plant fibers suggests attempts to strengthen or lighten the material. The result was a more versatile sealant or adhesive. These basic techniques laid an early foundation for more advanced refining methods that would appear in later centuries, although those developments would mostly come outside the scope of these two early civilizations.

Social and Political Dimensions

In regions where bitumen was abundant, control over seepage sites could become politically significant. A ruler who had easier access to this resource could build stronger city walls, more watertight canals, and maintain well-sealed boats. This advantage might translate into greater resilience against floods or invasions. While there are no records of major wars fought solely for control of bitumen in ancient Mesopotamia or Egypt, it is plausible that conflicts over resource-rich areas included the value of bitumen among other commodities.

For instance, the city of Hit in Mesopotamia was known for bitumen pits, and records suggest it was an important location for supplying pitch to Babylon and other major cities. Whoever governed Hit could trade or withhold bitumen to influence relations with neighboring polities. Similarly, Egyptian control over the Sinai or certain parts of the Levant might have been partially motivated by access to trade routes that delivered bitumen and other critical goods.

The Broader Impact on Technology

The advanced use of oil-based substances in Mesopotamia and Egypt hints at a broader pattern: whenever a resource proved useful, ancient civilizations found diverse ways to apply it. The skill sets developed around gathering and using bitumen—like boat sealing, building waterproof foundations, or preserving bodies—demonstrated a level of technological ingenuity often overlooked in discussions of the ancient world.

CHAPTER 3: OIL IN ANCIENT CHINA AND EAST ASIA

Introduction

In ancient times, different regions developed unique relationships with oil. While the Near East and North Africa saw early use of bitumen for construction and religious rites, China and East Asia approached oil in their own ways. This chapter looks at the role of oil in ancient China and surrounding areas, focusing on how people discovered it, extracted it, and integrated it into daily life. We will see that communities in China used oil for lighting, heating, and in some cases, for early industrial processes such as salt extraction. At the same time, other cultures across East Asia—like those in what is now Japan and Korea—had their own, albeit smaller-scale, experiences with natural seepages and possible oil uses.

As we examine China's historical records, we find mentions of substances that resemble crude oil and bitumen, sometimes referred to by different names. These references, along with archaeological clues, reveal that ancient East Asian societies understood the potential of oil, even if they did not harness it on as large a scale as modern industries. They also benefited from trade routes like the Silk Road, which brought information about minerals and organic materials from the West. By piecing these details together, we can see how oil's story in East Asia contributed to humanity's broader understanding of this valuable resource.

Early Chinese Awareness of Oil

China has one of the oldest continuous histories in the world, with archaeological sites that go back many thousands of years. While the earliest human settlements in China did not leave explicit records of oil usage, there is evidence that by the Shang and Zhou dynasties (second millennium BCE), people knew of certain flammable liquids that could come out of the ground. These were not always labeled as "oil" in the modern sense. Sometimes they were described as "burning water" or "inflammable water," or even occasionally "rock oil," indicating that the substance was found seeping from rocky terrain.

Written records from later periods, such as the Han dynasty (206 BCE – 220 CE), sometimes refer to thick, black liquids used for certain tasks. Though many references are brief, they suggest that people recognized the flammable qualities of the substance. In some areas of ancient China, natural gas also seeped to the surface, which might have been confused with liquid oil in certain accounts. Yet the concept of a black, oily substance that could be ignited was not foreign to the inhabitants of that time.

These early hints show that while oil may not have been central to ancient Chinese life, it was hardly unknown. People living near seepage sites likely experimented with it, testing if it could burn for light or serve as a waterproof sealant. Over time, knowledge of these properties spread to other regions, though not always in a formal, organized way. Small communities might have had their local "oil secrets" passed down through families or taught to trusted neighbors, long before any imperial authority took note of them.

The Significance of Brine Wells and Co-Produced Oil

One of the most interesting aspects of ancient Chinese history is the drilling for brine (saltwater) in places like Sichuan province. The pursuit of salt was a major economic driver in China for many centuries. Salt was critical for preserving food and for many other purposes, which made it valuable. To reach saltwater, Chinese artisans developed deep drilling techniques that were remarkably advanced for their time. Some wells are believed to have reached hundreds of meters below the surface, a feat that required sophisticated tools and methods.

In the process of drilling for brine, workers occasionally encountered pockets of oil and natural gas. Historical records suggest that they sometimes collected this oil for use as fuel to evaporate the brine into salt crystals. Thus, oil entered the economic equation as a byproduct of salt production. Rather than discarding it as an unwanted substance, they recognized that it could be burned and that it had potential value.

This dual extraction process—getting saltwater from deep wells and encountering oil or gas along the way—set China apart from many other ancient civilizations. While Mesopotamia and Egypt used bitumen from surface seeps, the Chinese developed techniques for deeper exploration, driven by their hunger for salt. Over time, these techniques evolved, and references to "oil wells" began to appear in local histories. Although these wells might have been primarily meant for salt, the fact that oil was collected at all indicates a growing understanding of the substance.

Tools and Techniques for Ancient Drilling

Chinese drilling technology relied on a system of bamboo pipes and percussion drilling. Workers would use a heavy iron bit, suspended by ropes or cables, and repeatedly lift and drop it to chip away at the ground. Water or mud was used to help remove debris from the borehole. Once they reached the brine or oil pocket, they inserted bamboo tubes to line the well, preventing collapse and ensuring a clear path for liquid to flow upwards.

Bamboo was plentiful, strong, and hollow, making it an excellent natural pipe material. By linking multiple bamboo segments, workers could create long conduits that stretched from the surface to the underground source. This allowed them to pump or bail out liquids, using manual labor or simple machinery. When natural gas accompanied the oil or brine, they sometimes channeled that gas through bamboo lines to use it as fuel for the evaporation of salt.

Though these processes were mainly for salt extraction, the concurrent discovery of oil meant that some well sites became the earliest examples of systematic oil gathering. This was not "drilling for oil" in the modern sense, but it was close enough that later historians recognized it as a precursor to modern oil exploitation. People in ancient China, especially in Sichuan, became familiar with the black liquid that appeared with the saltwater and learned how to manage it carefully so they could put it to practical use.

Using Oil for Lighting and Heating

Because oil burns brightly and produces significant heat, it made sense that once people in ancient China had access to it, they tested it for domestic uses. Oil lamps could be constructed out of simple clay or metal containers, with a wick dipped in oil. Animal fats and vegetable oils were more common fuels for lamps across many ancient cultures, but in regions near oil seepages, crude oil could also be used. It might have produced more smoke or odor than refined fuels, but it was still a viable source of light.

In some places, the surplus oil from salt wells was used in small furnaces or stoves for heating. This practice was not widespread because wood, charcoal, and coal were more readily available in many parts of China. However, where surface or shallow seepages were abundant, oil might become a supplementary or alternative fuel. Over time, local artisans learned to skim off lighter fractions of the oil or separate thicker residues to reduce smoking.

These early experiments laid the groundwork for future generations to explore better refining techniques. Although large-scale refining plants did not exist in ancient China, the repeated use of oil for lighting and heating suggested that people were aware of its different properties. They recognized that some types of oil burned cleaner than others, and that sometimes oil needed to be heated or filtered before it was suitable for indoor lamps. Such hands-on knowledge contributed to the gradual improvement of oil usage over centuries.

Rituals and Medicinal Beliefs

Beyond practical applications, ancient China also had traditions that ascribed special properties to unusual natural substances, including certain types of oil or tar. In some medical texts from the later dynasties, there are brief references to ointments or balms that included tar-like substances. While not every mention explicitly states "petroleum," it is plausible that crude oil was among the ingredients used for treating external wounds or skin conditions.

Likewise, in rituals meant to purify or ward off evil spirits, substances that produced a strong flame or smoke could hold symbolic power. Some local sects might have used oil lamps in temples or shrines, believing that the unique smell or brightness had spiritual significance. Though we lack detailed records of these practices, the general principle of burning unique substances during ceremonies appears in many ancient cultures, and China was no exception.

A few anecdotal accounts from historical records mention traveling healers or spiritual practitioners using black, viscous fluids in their remedies. Whether these fluids were pure crude oil or a mix of tar and herbs is not always clear. Still, the fact that they appear in the written record indicates that oil-based substances sometimes played a role in the broader tapestry of traditional Chinese healing practices.

Influence of the Silk Road and Exchange of Ideas

The Silk Road was a network of trade routes linking China to Central Asia, the Middle East, and eventually parts of Europe. While silk, spices, and precious metals were the primary goods exchanged, knowledge about natural resources traveled along these routes as well. Traders and scholars brought tales of bitumen pits in Mesopotamia, oil seeps in Persia, and even early uses of petroleum in places like the Caucasus region.

Chinese explorers or diplomats who ventured westward may have encountered more advanced oil usage or learned about different ways to refine or store the substance. Conversely, travelers from regions with established oil seepage sites might have shared techniques for harnessing oil in cooking, heating, or medicinal preparations upon arriving in China. Over many centuries, this cross-cultural dialogue enriched local understanding, even if oil never achieved the same central role in China that it later would in modern times.

Some historians suggest that the knowledge of distillation, which blossomed in the Islamic world, might have eventually reached China through these routes. While hard evidence is sparse, any introduction of basic distillation ideas would have enhanced local experiments with separating lighter and heavier fractions of crude oil. This step was crucial in discovering which parts of the oil burned best or had specialized uses.

Japan, Korea, and Other East Asian Regions

While China's large population and vast territory led to more extensive use of oil, smaller cultures in East Asia also encountered natural seepages. In Japan, certain regions are known for natural hot springs and occasional oil seeps, though in ancient times, these seepages were often minor. Historical records from Japan's Nara and Heian periods (8th to 12th centuries CE) mention unusual black liquids, sometimes used in small lamps or for waterproofing paper or wooden items. However, the limited availability and the island nation's reliance on other resources meant that oil did not become a widespread staple.

In Korea, references to bitumen or oil appear sporadically in texts from the Three Kingdoms period (1st century BCE – 7th century CE) and later Goryeo dynasty (10th to 14th century). These references are often connected to trade with China or the use of certain substances in temple rituals. Like Japan, Korea had other abundant resources, and large-scale oil exploitation never took off in the ancient era.

Despite the smaller scale of these uses, they confirm that oil was recognized in various parts of East Asia, not just in China. The interplay of trade, cultural exchange, and local resource availability shaped how each region perceived and integrated oil into its daily life or spiritual practices. Over time, pockets of knowledge spread across the area, ensuring that no single region remained entirely ignorant of oil's existence or potential uses.

Impact on Local Economies

Although China's economy was vast and diverse, regions that had salt wells co-producing oil could enjoy a modest advantage. The oil extracted might be sold to neighboring villages for lamps or for specialized crafts. It may have also reduced local reliance on animal or vegetable oils, which were often more expensive or needed for cooking. By burning crude oil in furnaces to evaporate salt brine, producers could lower some costs, thereby making salt production more profitable.

Yet, for most of ancient China, oil was not a primary economic driver. Agriculture, silk production, ceramics, and metalworking took precedence. Oil was supplementary—a helpful discovery that complemented existing industries. In times of war or unrest, control over brine wells (and by extension, oil) could still matter, but it was not on the scale of controlling farmland or major trade routes. So, while oil had an economic presence, it did not become the central commodity that would one day shape entire nations in the modern era.

Literary References and Cultural Perception

Chinese literature, especially from the Tang (618–907 CE) and Song (960–1279 CE) dynasties, sometimes mentions flammable liquids or "stone oils." Poets might refer to them in passing as curiosities or as part of a vivid description of a region's natural wonders. Official records or gazetteers compiled by local governments often included short notes about resources, such as an entry stating, "In this district, there is a spring that yields a black liquid, which is used for lamps."

These references show that while oil was not front-and-center in daily life, it was noteworthy enough to record. Authors or historians who were interested in nature, commerce, or geography would mention these sites as part of their effort to map the empire's resources. Over time, this accumulation of knowledge likely helped administrators plan taxes or manage resources more effectively. But it also gave future generations a window into the slow but steady growth of oil awareness throughout the region.

In many cases, oil was associated with wonder or mystery. A burning liquid was inherently impressive to cultures used to solid fuels like wood or charcoal. Even if it produced smoke or an unpleasant smell, the idea that the earth itself could yield a fiery substance stirred the imagination. Some texts describe peasants or travelers who stumbled upon oil seeps and were amazed that they could light the strange liquid on fire.

Health and Environmental Considerations

Though limited in scale, oil usage in ancient China and East Asia raised some early concerns or at least awareness about potential risks. For example, using crude oil for indoor lamps could produce strong fumes. People in poorly ventilated homes might have noticed headaches or irritation when burning oil for too long. In salt wells where oil and gas were present, the risk of fire or explosion was higher if there was a spark. Workers learned to handle bamboo pipes and drilling equipment carefully to prevent accidents.

On the environmental side, large-scale pollution was unlikely because the extraction levels were small and localized. Nevertheless, near brine wells, some local streams or soil might have been contaminated if oil was carelessly dumped or if a seep spilled over after heavy rain. While no organized environmental movement existed in ancient China, a basic sense of caution may have emerged whenever accidents occurred. Observing the black slick on a river's surface or the destruction of nearby vegetation could have led some communities to treat oil with a degree of respect and wariness.

Technological Overlaps and Inspiration

Ancient Chinese innovation is famous for inventions like paper, gunpowder, and the compass. Though oil was not credited with the same level of revolutionary impact, the drilling methods perfected for salt wells laid a foundation for future technological progress. The concept of deep drilling, bamboo piping, and controlling subterranean liquids would prove crucial if and when the demand for oil grew. Even centuries later, during other dynasties or in more modern periods (beyond our book's scope), those basic principles would continue to guide oil exploration.

Additionally, the presence of natural gas in many of these wells might have spurred curiosity about other underground resources. If black liquid and invisible gas could erupt from the depths, what else lay below? Such questions subtly encouraged a spirit of observation and experimentation. Over time, local craftsmen and scholars continued improving the processes for harnessing these geological treasures.

Comparisons to Western Ancient Uses

While this book focuses on the historical timeline up to the late 19th century and tries not to draw too many modern comparisons, it is still interesting to note that ancient Chinese oil usage developed independently of the Middle Eastern and Mediterranean traditions. Mesopotamia and Egypt had easy access to surface bitumen, while China's primary impetus for drilling was salt. These varied starting points led to distinct methods and traditions.

In the West, bitumen was widely used for waterproofing and building, whereas in China, brine wells and the potential for oil as a byproduct took the spotlight. Both traditions, however, recognized oil's flammable nature and possible applications in lighting and heating. They also showed that once people gained a foothold on how to gather oil, they slowly discovered new ways to put it to use. The universal lesson is that wherever a valuable or unusual resource appears, human curiosity and ingenuity tend to flourish around it.

Limitations and Scale

It is important to keep the scale of ancient Chinese and East Asian oil usage in perspective. Despite the documented evidence, this was far from the large-scale drilling operations we might imagine. The wells were often modest, the tools handmade, and the distribution network small. Only in certain localities did oil become significant enough to note in official records. Most farmers, artisans, and merchants relied on other fuels and materials for daily life.

Nevertheless, each small step contributed to a broader historical tapestry. A handful of wells producing oil alongside salt could still influence the local economy and technology. Over many centuries, incremental knowledge about how to handle, store, and burn this substance would accumulate. While no single ancient Chinese text stands out as a "petroleum manual," the collective wisdom embedded in scattered references and archaeological finds tells us that oil was never entirely absent from the story of East Asia.

Legacy for Future Eras

As dynasties rose and fell, and as China engaged with neighboring regions, the understanding of oil continued to evolve. The seeds planted during these early centuries would eventually germinate into more sophisticated uses. By the time Europeans arrived in East Asia in later centuries (well beyond our current focus), they sometimes marveled at Chinese drilling methods, which appeared advanced compared to some of their own practices.

From an even broader viewpoint, the ancient Chinese experience with oil reminds us that resource utilization is often a side effect of seeking something else. In this case, the pursuit of salt led to the discovery and partial exploitation of oil. This pattern—where oil emerges as an accidental find—repeats in other parts of the world, too. Ancient societies might have started with bitumen for building or shallow tar pits for occasional use, only to realize later the deeper potential of liquid crude or natural gas. In China, that realization was nurtured by a well-developed drilling tradition, which did not reach full bloom in antiquity but certainly left an important mark on history.

CHAPTER 4: OIL IN CLASSICAL GREECE AND ROME

Introduction

Classical Greece and Rome loom large in the popular imagination for their advances in philosophy, government, law, and engineering. Yet these civilizations also engaged with oil and bitumen, albeit in ways that might differ from the grand scale of their other achievements. In this chapter, we explore how the Greeks and Romans encountered and used oil-based substances, including pitch, tar, and crude oil seeps. We look at their military applications (such as the famous "Greek fire"), maritime endeavors, architectural feats, and daily life.

By delving into ancient Greek and Roman texts, archaeological findings, and historical accounts, we see that oil, while not a dominant resource, played a supporting role in various spheres. The wide reach of the Roman Empire meant that knowledge of oil seepages in different provinces traveled back to the center, feeding curiosity and occasional exploitation. Meanwhile, Greek city-states, known for their seafaring culture and trade networks, likely encountered oil-based materials during their extensive voyages. Overall, the story of oil in Classical Greece and Rome adds another layer to our understanding of how ancient peoples harnessed the earth's natural gifts.

Early Greek Encounters with Petroleum Substances

Greece's geography includes mountainous terrain, numerous islands, and a coastline that facilitated trade and travel. In antiquity, Greek sailors journeyed across the Mediterranean, occasionally coming into contact with areas where oil seeped from the ground or where bitumen could be collected. One such region might have been near the Ionian coast or in parts of the Aegean, where small seeps were reported in later historical records.

Some early Greek texts use terms that can be translated as "bitumen" or "naphtha," indicating a sticky or flammable substance. While references can be vague, they reveal that at least some Greek communities recognized oil's ability to burn and its potential as a sealant. As Greek colonies spread to places like Asia Minor and the Black Sea region, they might have observed local uses of bitumen, possibly influenced by neighboring cultures such as the Persians. These interactions enriched Greek knowledge about oil's properties, even if there was no systematic oil industry at the time.

Greek historians and natural philosophers—figures like Herodotus, Aristotle, and later writers—occasionally commented on natural wonders, including springs that produced flammable liquids. While these references are usually brief, they confirm that the Greeks knew about oil seepages. They also recognized that some springs or lakes had a layer of "pitch" on top that could be collected and used for specific tasks. Although overshadowed by their discoveries in mathematics, science, and medicine, the Greeks were not ignorant of petroleum-based substances and their potential uses.

Military and Maritime Applications

One of the more notable uses of oil or oil-based substances in ancient Greece and later in the Byzantine era (which is outside our direct scope but still relevant for context) was in military technology. The concept of "Greek fire" stands out, though it was more famously employed by the Byzantine Empire after the classical period. Still, the roots of such flammable weapons can be traced back to experiments with different combustible materials, possibly including crude oil, naphtha, and pitch.

In classical times, Greek navies occasionally used flaming projectiles—arrows or pots filled with flammable substances—to damage enemy ships. While this might not have been as advanced as the pressurized "Greek fire" of later centuries, the principle was the same: a combination of resin, bitumen, sulfur, or other ingredients that could stick to ships and continue burning on water. Thus, we see how knowledge of oil's flammability intersected with Greek innovations in warfare.

Maritime activities also required waterproofing, especially for wooden ships. Greeks relied on pitch or tar, derived from pine resin as well as from natural bitumen, to seal hulls and protect them from decay. Although pine resin might have been the most common ingredient, mixing in bits of bitumen was not unheard of if it was available. This combination created a strong protective coating that improved the ships' durability. Given Greece's dependence on naval power—both for trade and defense—any enhancement in ship maintenance had considerable value.

Philosophical and Scientific Observations

Greek thinkers did more than just record the existence of oil. Some, like Aristotle, tried to categorize natural substances and understand their properties. While the bulk of Aristotle's writings on geology or natural sciences focus on broader topics, his mention of "oily waters" or "earthy exhalations" hints that he was aware of phenomena resembling oil seeps. Other philosophers and physicians, such as Theophrastus, also made notes about geological curiosities, including places where the ground oozed tar or bitumen.

The Greek tradition of scientific inquiry encouraged observation and classification of the natural world. Though they did not have the same experimental approach seen in modern science, they laid the groundwork for rational examination. By naming substances like naphtha, describing their source, and noting their properties, they contributed to a growing body of knowledge about flammable liquids. This curiosity was not limited to academic circles; trade merchants and explorers brought back stories of unusual materials from distant lands, often fueling the interest of scholars who wanted to fit these discoveries into a broader worldview.

Oil in Daily Life and Small-Scale Industry

For most ordinary Greeks, oil would have had a minimal presence in everyday life compared to staples like olive oil. Yet, in certain regions with natural seepages, small communities might have gathered the substance for tasks such as sealing baskets or repairing roofs. The strong adhesive and waterproofing qualities of bitumen made it useful for patching leaky containers or protecting wooden structures from moisture. It was not as widespread or prized as building materials like stone or timber, but it found its niche.

Occasionally, bitumen might have been used in small-scale industries. Tanners or leather workers could employ it to waterproof hides. Potters and craftsmen might use it to create a unique finish on certain items. However, these applications were sporadic and limited by the availability of the material. Unlike Mesopotamia, Greece did not have large surface tar pits or major seepage sites that could supply a constant flow of bitumen. Instead, what existed was typically localized and moderate in output.

Because bitumen and crude oil-based substances could produce unpleasant odors and smoke when burned, their application in lighting or heating was less common in Greece. Olive oil was readily available for lamps, and charcoal or wood served as primary fuels for heating. Only in situations where a cheaper or alternative sealant was needed, or for specialized uses in warfare, would Greeks turn to petroleum-based materials.

The Roman Empire: Infrastructure and Resource Management

When we shift our focus to the Roman Empire, we encounter a civilization that excelled at large-scale infrastructure projects—roads, aqueducts, public buildings, and more. Romans prided themselves on their engineering prowess, and they also had an efficient administrative system to manage resources across a vast territory. As the empire expanded, it encompassed regions with known oil seepages, such as parts of North Africa, the Near East, and areas near the Caspian Sea (although that region was on the empire's periphery). Romans gathered information about local resources, including bitumen deposits, as they integrated these provinces into their domain.

One famous example of Roman interest in petroleum-based substances is the use of Dead Sea bitumen. Referred to as "Jewish pitch" in some historical accounts, the Dead Sea area was known for lumps of asphalt or bitumen washing ashore. This material was valuable for sealing, waterproofing, and even for medicinal recipes. Roman administrators might have regulated its collection or taxed it, seeing it as a commodity that could be shipped throughout the empire. Soldiers, builders, and merchants who traveled with the Roman legions or along trade routes likely spread knowledge of bitumen's applications to distant provinces.

Roman Construction and Waterproofing

Roman concrete (opus caementicium) is often praised for its durability, especially in structures like the Pantheon and numerous aqueducts. While this concrete typically relied on volcanic ash (pozzolana) and lime, bitumen sometimes appeared in specialized waterproofing tasks. For example, certain sections of baths or cisterns could be lined with a bitumen-lime mixture to prevent leakage. The Romans knew that water could erode mortar over time, and bitumen added an extra layer of protection.

Additionally, Roman roads, which spanned from Britain to the Middle East, required careful design to withstand heavy use and variable weather. Although stone and gravel were the primary materials, some engineers experimented with layering pitch or tar in certain spots prone to water damage. This might not have been widespread, but it shows that Roman ingenuity often took advantage of any material that could extend the life of their infrastructure.

Romans also built extensive maritime facilities—harbors, docks, and ships. Like the Greeks, they needed to seal wooden hulls. Pitch derived from pine or other trees was the standard. However, if a colony had nearby oil seepages, local shipbuilders might blend in some natural bitumen, creating a more robust sealant. Over centuries, these small-scale practices contributed to the empire's reputation for engineering excellence.

Medical and Cosmetic Uses in the Roman World

The Romans were famously interested in various balms, ointments, and cosmetics, some of which included exotic ingredients. Pliny the Elder, in his encyclopedic work "Natural History," mentions substances similar to bitumen for their possible medicinal properties. Certain texts note that applying pitch or tar could help close wounds or reduce infections, although modern science would question the safety and efficacy of such treatments.

In cosmetics, it was not uncommon to see references to black or dark substances used for hair dye or eyebrow enhancements, a trend possibly adopted from Eastern cultures.

These might sometimes have contained small amounts of tar-like material. However, given the strong smell and consistency, it is more likely that Romans preferred less pungent components if they could afford them. Still, the notion that oil-based materials could be used in limited medical or cosmetic contexts underscores how resourceful the empire was in experimenting with natural products.

Military Engines and Warfare in the Roman Context

As the Roman Empire expanded, its legions clashed with numerous adversaries, each employing different tactics and weaponry. The Romans learned from these encounters and also adapted foreign technologies. Like the Greeks, they were aware of flaming arrows or projectiles that could be launched using torsion siege engines or catapults. Bitumen or naphtha was sometimes used to enhance the fire's intensity, ensuring it stuck to enemy fortifications or even ships.

While the Romans did not develop a technology quite as iconic as "Greek fire," they still recognized that flammable liquids could serve as a psychological and tactical weapon. Some accounts describe the use of burning pitch-filled pots hurled over city walls during sieges. This approach, though brutal, was part of the era's warfare methods. Yet again, we see how oil or oil-related materials, despite being a minor aspect of Roman logistics, could play a crucial role in certain military operations.

Trade Networks and Resource Flow

Under Roman rule, trade networks stretched across the Mediterranean and beyond. The Pax Romana allowed merchants to move goods relatively safely. If a particular region had bitumen or some form of natural oil, they could send it to other parts of the empire, where it might fetch a good price among shipbuilders, engineers, or even apothecaries. These flows of resources expanded local knowledge: a craftsman in Gaul might learn about bitumen from Judea, or a merchant in Egypt might import pitch from the Black Sea area.

Ports like Ostia (near Rome) and Puteoli (in the Bay of Naples) were hubs where ships unloaded cargoes from across the empire. Among the countless amphorae of wine, olive oil, and grain, there might be occasional containers of pitch or lumps of asphalt. Detailed Roman customs records do not survive in full, but references in letters and legal texts confirm that the trade in these substances existed, albeit in much smaller quantities compared to staple goods.

Roman Attitudes Toward Natural Phenomena

Rome's educated classes, much like the Greeks, took an interest in documenting natural curiosities. Writers such as Pliny the Elder, Strabo, and Vitruvius wrote about geology, geography, and architecture, respectively, sometimes noting the presence of bituminous springs. They offered explanations, often mixing observation with philosophical reasoning. For instance, they might argue that heat within the earth caused certain regions to exude oily substances, or that a combination of sulfur and minerals gave rise to flammable liquids.

These ancient authors' curiosity paved the way for future understanding, even if their theories were incomplete or imprecise by modern standards. The mere act of cataloging facts about oily seeps or unusual lakes helped preserve knowledge, which scholars of later eras could revisit. In this sense, the Roman Empire functioned as a vast library of information, not just for law and politics but also for the natural sciences of the time.

Lesser-Known Provinces and Their Oil Legacy

While the heart of the Roman Empire was in Italy, the empire encompassed diverse lands with varying resources. Provinces in North Africa, such as what is now Tunisia or Libya, had scattered tar pits or asphalt deposits that local tribes might have used even before Roman rule. In the East, near the Caucasus, there were areas where natural flames burned above ground due to escaping gas, and these sometimes were accompanied by oily seepages. Romans stationed in these border provinces may have documented local customs, leading to a cross-pollination of ideas about how to harness these substances.

Archaeological digs in distant Roman outposts occasionally reveal traces of bitumen used in building materials or as a sealant for water cisterns, indicating that local garrisons or administrators employed whatever resources were available nearby. Although oil was not a top resource priority for the empire, the presence of Roman engineering in remote regions meant that knowledge and usage of bitumen inevitably spread. Over time, these small pockets of usage added up to a broader Roman familiarity with the substance.

Cultural Reception and Symbolism

In a cultural sense, black, tar-like substances might sometimes have carried symbolic weight in Roman religion. The Romans worshipped a pantheon of gods, and certain temples or altars might have used pitch or tar in ceremonies involving fire. Smoke and flame were potent symbols in many ancient rites, representing purification or the presence of divine power. Though we do not have detailed records of specific "oil rites," the general practice of burning substances for sacrificial purposes was widespread.

If a region's religious practices involved burning pitch as an offering, the Romans might incorporate or tolerate these rites in their governance, provided it did not conflict with the empire's basic expectations. In some provinces, local deities associated with natural phenomena—like hot springs or fiery vents—were folded into the Roman religious framework. This again shows the adaptability of Roman culture when it came to absorbing the beliefs and materials of conquered peoples.

Technical Limitations and Safety Concerns

While the Romans were advanced for their time, they still faced technical limitations when dealing with oil or bitumen. They lacked drilling technology on par with what ancient China had for salt wells, and much of their oil-based material came from surface seeps or easily accessible deposits. Transporting thick, sticky substances was cumbersome, often requiring containers lined with pottery or leather. The risk of spills or accidental fires was present whenever flammable materials were moved long distances.

Additionally, stored oil or bitumen could degrade or become contaminated with dirt, water, or plant matter. Without refined processes for filtering or distillation, the quality of the material varied. Some shipments arrived in lumps that had to be melted or softened before use. Others might have been partially diluted with other substances to increase volume. These realities limited the scale at which Romans could adopt oil for everyday use, reinforcing its place as a niche resource rather than a cornerstone of the economy.

Transition and the Later Empire

As the Roman Empire transitioned into its later phases, eventually splitting into Eastern and Western entities, many practices evolved. The Eastern Roman (Byzantine) Empire, centered in Constantinople, famously developed a more sophisticated form of "Greek fire" in the 7th century CE—beyond our current historical cutoff but still worth a brief mention. This maritime weapon relied heavily on petroleum-based substances, indicating that knowledge of oil's flammable properties was preserved and even enhanced in the Eastern Roman world.

In the Western half, invasions, economic downturns, and social upheavals disrupted trade networks. Many specialized knowledge areas, including advanced engineering practices, suffered setbacks. While some pockets of expertise lingered on in monastic communities or local traditions, the large-scale Roman approach to resource management waned. Nevertheless, the legacy of Roman usage of oil-based substances did not vanish entirely. It fed into medieval Europe's scattered knowledge base, eventually blending with Middle Eastern developments that reentered Europe during the Crusades and later periods.

Conclusion of Chapter 4

Classical Greece and Rome may not have elevated oil to the same central position as they did marble, olive oil, wine, or precious metals, but they certainly recognized the value of petroleum-based substances in specific contexts. From sealing ships and fortifications to fueling early incendiary weapons, oil and bitumen served practical and military needs. Philosophers and historians documented these resources, ensuring that future generations would know of "fiery springs" and "sticky tar pits."

In examining the ancient Greeks and Romans, we see civilizations that were resourceful enough to exploit local seepages and trade networks, yet constrained by the limited technology for extracting oil from deeper underground. Their experiences formed another important link in the global chain of oil history, bridging the developments in the Near East with those that would occur later in Europe and elsewhere.

Now that we have explored the uses of oil in the classical Mediterranean world, we will move on in subsequent chapters to look at how oil's story continued through the early Islamic era, medieval Europe, and beyond. Each region and era added its own creativity, shaping the evolving narrative of how humans discovered, traded, and utilized this remarkable resource.

CHAPTER 5: OIL IN THE EARLY ISLAMIC WORLD

Introduction

The early Islamic world, spanning the 7th century onward, witnessed the rapid spread of a new civilization across vast regions—from the Arabian Peninsula to parts of Africa, Persia, Central Asia, and beyond. This expansion was not just political or military; it also brought about significant cultural, scientific, and technological changes. Among many developments was the evolving use and understanding of natural resources like oil.

In this chapter, we will explore how the early Islamic world approached petroleum-based substances, focusing on several key themes: the use of oil in daily life, the role of trade in spreading knowledge, advancements in distillation and alchemy, and the influence of religious and cultural factors on resource utilization. While oil was not the dominant resource of this era—compared to items like textiles, spices, or precious metals—it nonetheless found its place in the bustling economies and innovative scientific circles of the time.

We will look closely at how scholars, traders, and craftsmen contributed to oil's gradual integration into various practices. Along the way, we will see that the expansion of Islamic rule created a network of communication linking lands that contained rich oil seepages. This laid the groundwork for more organized extraction and experimentation. Although this period did not witness the industrial-scale drilling of the modern age, it did establish core ideas and techniques that would prove crucial in later centuries.

1. The Historical Context of Early Islam and Resource Management

The founding of Islam in the 7th century ushered in a new era of cultural and political unity in the Middle East and beyond. Beginning in the Arabian Peninsula, Islamic rule soon spread to places like Syria, Egypt, Persia (modern-day Iran), parts of Central Asia, and across North Africa. As these regions fell under one overarching political sphere, trade and knowledge exchange became more fluid.

Merchants, soldiers, and scholars moved relatively freely through this sprawling territory, sharing ideas about agriculture, architecture, medicine, and more. With the consolidation of regions previously governed by the Byzantine Empire, the Sassanian Empire, and various local rulers, knowledge about natural resources—including oil—merged into a broader body of collective expertise. Areas known for natural bitumen or oil seepages, such as regions in modern Iraq, Iran, and the shores near the Caspian Sea, came under Islamic governance.

Islamic leaders, particularly during the Umayyad (661–750 CE) and Abbasid (750–1258 CE) dynasties, were often keen to support technological and economic innovation. They recognized that prosperity in agriculture, trade, and manufacturing would strengthen their rule. While water management, agricultural improvement, and long-distance commerce took priority, interest in any potentially useful resource—like oil—fit into this bigger picture of harnessing nature's gifts for the benefit of the growing empire.

2. Early Uses of Oil in Daily Life

Many communities within Islamic territories had long histories of using bitumen or crude oil. In areas like Mesopotamia (modern Iraq), the tradition stretched back thousands of years. Under the new political framework, these existing practices continued and sometimes evolved. For common folk, oil's main uses remained relatively straightforward:

1. **Waterproofing and Construction**: Similar to how older Mesopotamian civilizations used bitumen for sealing and waterproofing, early Islamic builders in these regions often continued applying it to walls, foundations, and roofs, especially in areas prone to flooding.
2. **Lighting**: Although lamps fueled by animal fats or plant oils were still common, in places where naturally occurring liquid oil seeped, it could be collected and used for lighting. Crude oil might have produced a strong smell and smoke, so it was not universally favored. But for those with limited access to other fuels, it offered an alternative.
3. **Medicinal Preparations**: Local healers sometimes incorporated tar or bitumen into ointments, believing them to have antiseptic properties. Although not systematically studied in a modern sense, these uses were part of broader traditional medicine.
4. **Adhesives and Sealants**: Beyond construction, bitumen could act as a glue-like substance. Craftsmen who built containers, water vessels, or even simple boats might have used it to reduce leaks.

Not everyone had access to these resources, of course. In desert regions like the Arabian Peninsula, bitumen seepages were less common, so people there relied on the knowledge and supplies imported from other provinces. Over time, local practices in oil-rich areas merged with broader Islamic cultural and economic systems, ensuring that news of oil's usefulness spread farther than ever before.

3. The Expansion of Trade Routes

One of the major achievements of the early Islamic world was the expansion and revitalization of trade routes connecting East and West. Caravans traveled across land routes like the Silk Road, while ships crossed the Red Sea, the Persian Gulf, and the Indian Ocean. Many of these routes passed through or near regions where oil seeped to the surface, such as the shores along the Persian Gulf or areas near the Caspian.

These routes served more than just the exchange of goods like spices, textiles, and metals. They also facilitated the movement of ideas. Traders and scholars from Persia might share knowledge of advanced distillation techniques with counterparts in Arabia or Egypt. Conversely, travelers from North Africa or the Levant could bring back tales of natural asphalt found in Mesopotamia. This network of information gradually helped unify and expand understanding of oil's properties and potential uses.

At certain caravan stops or port cities, markets offered lumps of bitumen or small quantities of crude oil for sale. While these products were not as high-profile as precious spices or silk, they found a consistent niche among buyers needing waterproofing materials or substances for lamp fuel. Over generations, these humble transactions ensured that oil became a familiar commodity in many corners of the Islamic world.

4. Advances in Distillation and Alchemy

A significant leap in the understanding of oil occurred due to the Islamic Golden Age's focus on science, medicine, and alchemy (an early form of chemistry). Between the 8th and 13th centuries, scholars worked in major intellectual centers like Baghdad, Damascus, and Cordoba, collecting and expanding upon knowledge inherited from Greek, Roman, Persian, and Indian sources.

4.1. The House of Wisdom in Baghdad

The Abbasid caliphate established Baghdad as its capital in the mid-8th century. This city became a vibrant hub of scholarship, hosting institutions like the Bayt al-Hikmah (House of Wisdom). Scholars there translated key texts on mathematics, astronomy, medicine, and alchemy. Figures like Al-Kindi (Alkindus), Al-Razi (Rhazes), and later Ibn Sina (Avicenna) made notable contributions to scientific thought.

While much of their attention centered on medicine, astronomy, or philosophy, some scholars studied practical chemistry. They devised equipment such as alembics and retorts that could be used to distill substances more accurately than before. Among the items subjected to distillation were petroleum-based materials, though references to systematic crude oil refining are scattered and less explicit than studies of, say, perfumes or medicines.

4.2. Al-Razi (Rhazes) and Petroleum Substances

Al-Razi, a 9th–10th-century Persian physician and alchemist, is famous for his works on medicine and chemistry. In certain treatises, he mentioned substances resembling naphtha and described processes for refining or separating lighter fractions from heavier oils. His intent often centered on creating medical ointments or better lamp fuels, but in doing so, he provided an early blueprint for how to manipulate crude petroleum.

Al-Razi's experiments hinted at the possibility that oil could be separated into components with distinct properties. While this did not lead to large-scale refining plants in his day, it did spark curiosity among subsequent scholars. Alchemists recognized that "naphtha," a more volatile and flammable fraction, could be harnessed for special purposes, including incendiary weapons (though these are more famously associated with the later Byzantine use of "Greek fire").

4.3. Ibn Sina (Avicenna) and Beyond

Ibn Sina, a towering figure in medieval medicine and philosophy, also contributed to general scientific knowledge. Though his works do not revolve entirely around petroleum, occasional remarks in his writings suggest that he was aware of its role in healing and as a combustible material. Other scholars, in the centuries that followed, continued incremental improvements in distillation apparatus.

These scientific endeavors did not create an "oil industry" as we would understand today, but they enriched the collective understanding of oil's potential. They showed that by applying heat and clever apparatus, one could modify the properties of crude substances. Over time, this knowledge would trickle down into practical applications, helping craftsmen produce slightly more refined lamp oils or specialized chemical mixtures for various tasks.

5. Incendiary Weapons and Military Uses

The early Islamic world, like many civilizations of the time, recognized that flammable substances could be used as weapons. Although the Byzantine Empire's "Greek fire" is the most famous example of medieval incendiaries, Islamic armies also experimented with mixtures containing naphtha, bitumen, and other flammable components.

Historical accounts describe siege warfare where pots or barrels filled with a combustible blend were catapulted over walls. Soldiers used hand-thrown fire grenades that combined oil-based substances, sulfur, and quicklime. In some cases, these weapons were used defensively to destroy siege engines or repel enemy forces attempting to scale walls or cross moats.

While the scale of such activities varied, they demonstrate that knowledge of oil's flammable qualities spread well beyond local lamps or adhesives. Military engineers in places like Syria, Persia, or North Africa used these mixtures to bolster the defenses of fortified cities. Commanders who valued such technology ensured a steady supply of the necessary materials, thereby encouraging a modest but reliable trade in petroleum-based substances.

6. Religious and Cultural Considerations

Islam, as a religion, did not forbid the use of oil or bitumen. However, the handling of flammable or foul-smelling materials sometimes came under scrutiny if it disrupted community life or interfered with the cleanliness required for prayers. Communities often set aside specific areas for tasks involving strong odors—like tanneries or bitumen melting—to avoid contaminating mosques or residential districts.

On the other hand, certain religious architectural projects could benefit from bitumen. Large mosques or madrasas (schools) sometimes needed waterproofing in their foundations. Builders who borrowed techniques from older Mesopotamian or Persian styles might incorporate bitumen layers to safeguard against groundwater. Over time, these practices blended into the broader Islamic architectural heritage without causing conflict with religious rules, as long as cleanliness and safety were maintained.

Cultural traditions also played a role. For instance, local folklore in areas with active seepages might have attributed spiritual significance to places where the earth leaked black tar or flammable gas. With the spread of Islam, some of these sites might gain new interpretations—perhaps seen as signs of the earth's hidden riches or simply natural wonders created by God. Whatever the perspective, there was no widespread taboo against collecting or using oil; it remained a neutral resource, shaped mainly by practical and economic considerations.

7. The Persian Gulf and Caspian Sea Regions: Hubs of Petroleum Activity

Among the most notable areas for oil seepage in the early Islamic world were regions around the Persian Gulf and near the Caspian Sea. Modern-day Iran, Iraq, and parts of Azerbaijan all experienced natural leaks of petroleum. Some local populations had been aware of these seepages since antiquity, but under Islamic rule, these regions became integrated into larger trade and political networks.

- **Persian Gulf**: Ports along the Gulf served as gateways for maritime trade linking Arabia, Persia, East Africa, and South Asia. While merchants focused primarily on pearls, dates, and other commodities, they also encountered oil seeps along the coastline. Over time, certain local entrepreneurs recognized the value of selling bitumen to passing ships for hull repairs.

- **Caspian Sea**: The area around Baku (in present-day Azerbaijan) was known for its "eternal fires," where natural gas seeped from the ground and ignited. Local dwellers also found crude oil seeping in shallow pits. Early Islamic geographers wrote about these marvels, noting that the region produced "naphtha" of high quality. Caravans traveling through the Caucasus might carry small amounts of this prized substance back south to cities in Persia or Mesopotamia.

These regional specifics highlight how geography and empire-building went hand in hand. Once part of the Islamic world, these territories found new markets for resources like oil. In turn, consumers in urban centers like Baghdad or Isfahan learned more about petroleum's range of uses. The result was a steady, if modest, rise in the collection and distribution of crude oil, naphtha, and bitumen, setting a precedent for future developments.

8. Integration into Urban Life

During the Islamic Golden Age, cities like Baghdad, Basra, Kufa, and others in the Fertile Crescent expanded rapidly. They boasted grand palaces, mosques, libraries, and bustling markets. Brick and stone construction remained dominant, but bitumen found its way into various niche uses:

1. **City Walls and Canals**: Large urban centers often had fortified walls, and some used bitumen to reinforce sections vulnerable to water infiltration. In cities along rivers, canals might be lined with bitumen to minimize leakage.
2. **Public Baths (Hammams)**: Hammams were integral to Islamic urban culture, serving hygiene and social purposes. While they mostly relied on firewood or other fuels to heat water, some references suggest that oil-based substances could be substituted in regions where they were cheaper or easier to obtain.
3. **Street Lighting (in Rare Instances)**: Most Islamic cities relied on simpler fuels for street lamps, and widespread street lighting as we know it did not exist in medieval times. However, wealthier quarters or notable public areas could occasionally use lamps filled with more refined oil. Though these cases were exceptions, they indicate experimentation with petroleum-based lighting in urban settings.

The overall impact of oil in daily urban life remained limited compared to major economic drivers like agriculture or textile production. Nevertheless, oil was present enough to be recognized, regulated by local authorities, and occasionally taxed or supervised to ensure safety and consistency in quality.

9. Regulatory Measures and Safety

Since flammable materials posed fire hazards, some cities instituted basic rules about storing and handling oil or bitumen. Shopkeepers dealing in these substances might be required to keep them away from open flames or residential areas. Guilds that oversaw various trades—like smiths, carpenters, or builders—might also have sub-guilds or sections responsible for the proper use of pitch and tar.

In some cases, local rulers or governors established designated zones for collecting bitumen from natural seepages. These zones could be taxed or leased, allowing the government to earn revenue and control distribution. While records of comprehensive petroleum laws are scant, the basic principle of resource management—ensuring supply, quality, and safety—was certainly present.

If an accidental fire involving oil broke out, local officials often blamed improper storage or negligence. This occasionally led to new ordinances or stricter oversight. The memory of large city fires in other cultures, such as those in Rome or Constantinople, likely influenced Islamic authorities to remain vigilant about potential conflagrations.

10. Scholarly Debates and Observations

While many references to oil in the early Islamic world are practical, some scholars engaged in philosophical or scientific speculation. They questioned how oil formed within the earth, whether it had medicinal properties, and how it related to other substances like water or metals.

Geographers and travelers—such as Ibn Hawqal, Al-Mas'udi, and Al-Idrisi—compiled extensive accounts of the lands under Islamic rule. These works sometimes mentioned oil seepages as local curiosities, along with descriptions of local customs. Although these travelers were more focused on topics like climate, agriculture, or cultural practices, their notes on flammable springs or tar pits provided valuable snapshots of oil usage in different provinces.

For example, Al-Mas'udi (10th century) traveled widely and wrote about various natural phenomena. While not devoting entire treatises to oil, he offered enough detail to show that by his time, people in certain regions recognized oil's commercial value. His writings, along with those of others, gave later historians glimpses into how integrated oil had become in everyday resource management.

11. The Influence of Interfaith Exchanges

During the early Islamic centuries, the empire included a diverse population: Christians, Jews, Zoroastrians, and others. These communities sometimes preserved older traditions of oil usage that predated Islamic conquests. Jewish communities in Mesopotamia, for instance, carried on older Babylonian practices involving bitumen, while Christian monasteries might have used certain forms of tar for preserving manuscripts or sealing roofs.

Furthermore, as Muslim, Christian, and Jewish scholars interacted—often through institutions of learning or shared commercial interests—they exchanged knowledge about everything from medicine to resource extraction. This interfaith dialogue contributed to a broader, more inclusive understanding of oil's properties. Over time, they also encountered ideas from beyond the Islamic world, including remnants of Greek and Roman texts discussing bitumen.

In certain border regions—like those abutting the Byzantine Empire—conflicts could erupt, but trade and knowledge exchange often persisted in peaceful periods. The synergy of multiple cultures living under relatively unified political structures ensured that local traditions relating to oil did not vanish; instead, they blended into a more diverse tapestry of practices.

12. Artisans and Craft Guilds

In many Islamic cities, craftsmanship was organized into guilds or associations that maintained standards and passed down trade secrets. While oil was not the main focus of most guilds, some specialized in waterproofing, boat-building, or construction tasks requiring bitumen or tar-like materials. These artisans developed practical methods for applying heated bitumen, blending it with other substances, and storing it safely.

Guild records from certain cities (though fragmentary) hint that masters would teach apprentices how to choose quality raw materials, handle them without causing fires, and refine them for particular tasks. The secrecy around some of these methods suggests a competitive market: a builder who knew how to make a more durable bitumen mixture could attract more clients.

Though overshadowed by luxurious crafts like metalworking or textiles, the "oil trades" still had their place. Over time, knowledge from these craftspeople spread along trade routes or moved with migrating workers. This grassroots level of technological exchange ensured that even distant communities might eventually learn about improved ways to handle petroleum-based substances.

13. Minor Refinement Efforts for Lamp Fuel

While large-scale refining was not an established industry, smaller-scale refinement did occur, especially for creating lamp oils with less smoke. Craftsmen might use simple distillation setups inherited from alchemical experiments, heating crude oil slowly to separate lighter fractions. This fraction—closer to what we might call kerosene—would be used in lamps, reducing unpleasant odors and soot.

Such efforts were likely limited to wealthier urban areas where there was a steady demand for better-quality lighting. Commoners generally relied on less refined substances or alternative fuels. However, these experiments demonstrated that the underlying principles of refining—applying heat and condensing vapors—were known and practiced, even if not widely commercialized.

Geographic proximity to seepages also played a role. In towns near the Caspian Sea or parts of Persia with abundant oil, local entrepreneurs had more incentive to refine and sell this resource. By contrast, in regions without easy access to petroleum, such as much of North Africa, the impetus to develop refining techniques was lower.

14. Environmental and Health Observations

Compared to modern times, the scale of oil usage in the early Islamic world was modest, so widespread environmental damage did not occur. However, local issues did arise:

- **Water Contamination**: People collecting oil from seepages near rivers or canals might unintentionally spill residue, affecting local water quality. Because water management was crucial in arid regions, any pollution that harmed crops or livestock drew the attention of local officials.
- **Air Quality and Smoke**: Burning crude oil in lamps or furnaces could produce thick smoke, raising health concerns or simply causing nuisance. City dwellers complained about the smell if refining or large-scale heating took place in residential areas.
- **Fire Hazards**: Improperly stored oil or bitumen led to accidents. Workshops caught fire if an open flame came into contact with spilled fuel. Over time, experience taught communities to keep flammable materials away from crowded bazaars and living quarters.

Though systematic environmental regulations were not as developed as today, local authorities did step in when problems threatened public health or safety. This created a rudimentary form of oversight that balanced economic benefit with communal well-being.

15. Decline and Transition

The early Islamic world thrived for several centuries, but like all historical periods, it eventually faced challenges. Internal political strife, the rise of rival dynasties, and invasions (such as the Mongols in the 13th century) disrupted established trade routes and scientific institutions. Cities like Baghdad suffered destruction, leading to a decline in scholarly output for a time.

Despite these upheavals, the knowledge about oil did not vanish. It was preserved in manuscripts, taught in smaller academic circles, and continued in practice by local communities. Even as political power fragmented, the overall cultural sphere of the Islamic world remained. In places like Egypt, Syria, or Persia, smaller regional powers carried on the tradition of resource management and craft production.

Later periods, especially those edging into the Renaissance in Europe or new empires in Asia, saw the next phase of oil's story. But within the early Islamic context, oil usage stayed at a moderate level—neither neglected nor becoming the prime driver of economies. It was part of the broader mosaic of resources that supported a vibrant, innovative civilization.

16. Legacy and Influence on Future Developments

The early Islamic contributions to understanding petroleum were subtle yet foundational. They included:

1. **Refinement Knowledge**: Alchemical experiments and small-scale distillation provided a glimpse into how oil could be separated into different components. This knowledge would later inspire more advanced refiners in other places and times.
2. **Geographical Integration**: By uniting oil-rich territories under a shared governance, the Islamic world made it easier for ideas and materials to move. This laid groundwork for the idea that oil was a tradeable resource, even if not on a massive scale.
3. **Scientific Preservation**: Many classical texts from the Greek and Roman eras discussing bitumen and flammable materials were preserved in Arabic translations. This allowed future scholars—both in the Islamic world and in Europe (after translations back into Latin)—to build on earlier observations.
4. **Cultural Acceptance**: Oil was integrated into Islamic architecture, medicine, and warfare without major religious or cultural prohibitions. This open-minded attitude meant that, when needed, communities could adapt and use petroleum-based substances efficiently.

By the end of the medieval period, many of these insights found their way into neighboring regions. Whether through the Crusades, expanding trade routes, or continued scholarly exchange, Europe and other parts of the world gradually absorbed the knowledge accumulated in the Islamic lands.

Conclusion of Chapter 5

In the early Islamic world, oil played a notable, if not dominant, role in daily life, scientific exploration, and economic exchange. Building on the legacies of ancient Mesopotamia, Persia, and other civilizations, Islamic scholars and craftsmen refined the usage of petroleum-based substances in small yet meaningful ways. They explored basic distillation, harnessed oil for warfare, and integrated bitumen into construction and waterproofing tasks.

Trade networks spanning the Middle East, North Africa, and parts of Asia spread familiarity with oil across diverse cultures. Religious considerations influenced how and where oil was handled, but no significant taboo prevented its use. Meanwhile, scientific inquiry—especially within centers like Baghdad—laid a foundation for more advanced chemical processes in later centuries.

Though the region faced political fragmentation and invasions over time, the knowledge and applications of oil did not disappear. Instead, they evolved, passed on through manuscripts, guild traditions, and trade secrets. As we move into the next chapter, focusing on medieval Europe, we will see how some of this Islamic knowledge crossed into European hands and shaped that region's slow but steady adoption of oil-based substances. In this way, the early Islamic world served as both a repository and a conduit for a resource that would eventually become central to global industrial development.

CHAPTER 6: OIL ACROSS MEDIEVAL EUROPE

Introduction

Medieval Europe, spanning roughly from the 5th to the late 15th century, was a time of great social, political, and economic change. Following the collapse of the Western Roman Empire, many European regions became fragmented into small kingdoms and feudal states. Throughout much of this period, daily life was shaped by agrarian economies, manorial systems, and the power of the Church. Although monumental engineering feats or large-scale trade networks had diminished compared to Roman times, knowledge continued to circulate, often via monasteries, traveling merchants, and, later, through interactions with the Islamic world.

In this chapter, we examine how Europeans of the Middle Ages encountered and utilized petroleum-based substances. Unlike the ancient Romans or the contemporary Islamic civilizations, medieval Europeans did not have wide access to oil-rich territories or a strong tradition of using bitumen. Still, they found certain ways to incorporate oil seepages and pitch into building, warfare, and religious life. Over time, contact with the Islamic world—particularly during the Crusades—reintroduced advanced techniques and sparked renewed curiosity in natural resources, including oil. By looking at Europe's medieval experience, we see how a region once dominant under Rome slowly rediscovered and adapted knowledge of petroleum that had largely remained dormant or localized.

1. The Early Middle Ages and the Fragmentation of Knowledge

The centuries immediately following the fall of the Western Roman Empire (5th to 8th centuries) were marked by political turmoil and reduced urbanization. Many public works, including roads, aqueducts, and large buildings, fell into disrepair. As a result, extensive uses of bitumen in construction declined, especially in regions that lacked easy access to local seepages.

Literacy rates dropped, and scholarly pursuits were often confined to monastic communities. These monasteries preserved some classical texts, potentially including references to bitumen or oil-based materials from authors like Pliny the Elder or Vitruvius. However, such subjects were not major priorities compared to theological writings, biblical commentary, or basic medicine. As a result, the knowledge of advanced engineering or resource management from the Roman era became fragmented, passed down in scattered manuscripts or local folk traditions.

In everyday life, people relied on wood, charcoal, and animal fats for heating and lighting. Where small deposits of pitch or tar existed—often derived from pine trees rather than petroleum—villages would use them to seal barrels, boats, and rooftops. Instances of actual crude oil usage during this early medieval period were sporadic, typically limited to areas with known seepages.

2. Local Oil Seepages in Europe

While Europe is not as famous for extensive oil reserves as the Middle East or certain parts of Asia, it still has regions where oil seeps naturally to the surface. Parts of what are now Italy, Germany, France, and even the British Isles contain small pockets of hydrocarbons that can leak or pool. During the Middle Ages, local populations near these seepages might have used the substances for:

1. **Waterproofing**: Coating the bottoms of boats or the roofs of simple huts, though this was less common than using plant-based pitch.
2. **Primitive Medicinal Applications**: Some communities believed the black, tar-like substance had healing or protective qualities for external wounds or skin conditions.
3. **Lamp Fuel**: Rarely used, except when there was no alternative. Crude oil tended to smoke heavily and release unpleasant odors, making it less appealing than animal fat or vegetable oil.

These uses remained localized, often overshadowed by more reliable materials. Since trade networks were weaker in early medieval Europe compared to the ancient Roman period, oil from one region rarely traveled far. Thus, knowledge about these seepages typically stayed within close-knit communities, passed down through oral tradition rather than widespread written treatises.

3. The Influence of the Church and Monastic Learning

The medieval Church played a central role in preserving knowledge, as monasteries became centers for reading, copying, and studying manuscripts. While theological texts dominated, some monks had interest in natural sciences, geography, or even alchemy. A handful of these monastic scribes might have encountered references to bitumen in older Latin works, possibly sparking minor curiosity about where such substances could be found in Europe.

Church-led construction projects, such as cathedrals and abbeys, were major undertakings. Stone was the primary material for these buildings, supplemented by mortar and sometimes lead for roofing. There is little evidence that petroleum-based substances featured significantly in cathedral construction. However, monastic scribes who read about Roman or early Christian uses of pitch for sealing catacombs or protecting wooden beams might have experimented with local materials if available.

The Church also required large amounts of candle wax for services. In places where wax or olive oil was expensive, a few communities may have turned to cheaper, heavier oils—though typically from animals or plants rather than crude petroleum. Overall, the role of the Church in sustaining interest in oil was minimal, but it did keep alive some fragments of classical knowledge in an otherwise largely agrarian and localized society.

4. Trade and Exchange Before the Crusades

Despite political fragmentation, some trade continued across medieval Europe. Overland routes connected northern regions to the Mediterranean, and maritime routes linked cities on the coasts. Italian city-states like Venice, Genoa, and Amalfi became hubs for commerce, facilitating the movement of goods from the Byzantine Empire and Islamic lands into Western Europe.

During these early centuries (roughly 8th to 11th), the majority of trade centered on high-value items such as spices, silks, precious metals, and luxury textiles. Although pitch and tar were traded, they were often derived from wood or other local resources rather than imported petroleum. Europeans at this time seemed unaware of the more sophisticated refining or distillation techniques being developed or preserved in the Islamic world.

Still, occasional travelers—scholars, pilgrims, or adventurers—may have brought back stories of burning springs or tar pits encountered abroad. These tales, often recorded in diaries or told around hearth fires, kept alive the notion that beyond Europe's borders, other cultures had found interesting ways to use flammable substances. Yet there was no large-scale impetus within Europe to explore or extract oil. The feudal economy, primarily focused on agriculture and local lords' estates, did not push for resource exploitation beyond immediate needs.

5. The Crusades and the Reintroduction of Eastern Knowledge

A major turning point came with the Crusades (11th to 13th centuries), when European armies traveled to the Holy Land. This clash and interaction with the Islamic world led to cultural exchange, albeit under often violent circumstances. Crusaders encountered advanced fortifications, siege tactics, and in some cases, the use of incendiary weapons. Islamic forces sometimes employed mixtures containing naphtha, learned from or parallel to Byzantine traditions of "Greek fire."

Returning crusaders not only brought back stories of these formidable weapons, but also impressions of Middle Eastern trade and resources. Although it would be an overstatement to say that the Crusades sparked a European oil rush, they did encourage more curiosity about substances like bitumen and naphtha. Some knights or merchants who remained in the Levant for extended periods might have witnessed how local craftsmen used crude oil in lamps or for water-resistant coatings.

Moreover, the increased flow of goods and ideas between East and West meant that texts on alchemy or natural philosophy—translated from Arabic into Latin—circulated more widely in Europe. These works contained references to distillation apparatus, chemical processes, and the potential of flammable liquids. Monasteries, universities (which were emerging in places like Paris and Bologna), and scholarly circles began to absorb and adapt these Eastern insights.

6. Alchemical Pursuits and Early Experiments

With renewed access to translated works from the Islamic world, European alchemists in the late medieval period found fresh inspiration. While alchemy focused largely on the search for the "philosopher's stone" and the transmutation of metals, many alchemists experimented with distillation. They studied various substances—herbs, metals, and occasionally petroleum—to discover their essence or unlock new properties.

European alchemists like Albertus Magnus or Roger Bacon may have read about Arabic distillation methods that could, in principle, separate crude oil into lighter fractions. While references to such attempts in surviving texts are rare, the general knowledge that oils could be heated and refined was undoubtedly transmitted.

However, large-scale practical application remained distant. Alchemical pursuits were often private, secretive, and not focused on commercial exploitation of oil. Furthermore, the scarcity of local seepages in Europe meant that even if an alchemist successfully distilled a portion of petroleum, there was not a robust network for widespread distribution or use. Still, these experiments laid a small foundation. They showed that certain fractions of oil could burn cleaner than raw crude, hinting at future possibilities.

7. Military Innovations and the Legacy of Incendiary Substances

Medieval Europe had a long history of using "Greek fire"-type weapons. Even before the Crusades, the Byzantine Empire (which many Europeans considered part of Christendom, even though culturally and politically distinct) had famously guarded the secret of Greek fire. This weapon relied on a mixture of petroleum, pitch, sulfur, and other ingredients to create a burning liquid that stuck to enemy ships.

Post-Crusades, some European armies attempted to replicate these incendiary formulas. While the exact recipes often remained elusive, the concept spread that combining a flammable oil-like substance with certain chemicals could yield powerful weapons. Siege warfare in the late Middle Ages did see the limited use of "fire pots" or "tar barrels" launched via catapults. If local sources of oil or bitumen were unavailable, pine tar or resin was substituted.

These innovations, though small in scale, reinforced the notion that if Europeans could access more consistent supplies of petroleum-based materials, they could potentially enhance their arsenals. But again, the economic and social systems of feudal Europe did not prioritize systematic resource extraction, so these ideas remained on the fringes.

8. Maritime Applications in the Late Middle Ages

Europe's maritime powers, especially in Italy, Spain, and Portugal, evolved significantly toward the end of the medieval period. Before the Age of Exploration truly began (which is more a 15th–16th-century phenomenon), shipbuilders sought ways to strengthen and waterproof their vessels. Traditional pitch and tar from pines or other trees were standard, but in certain ports, small amounts of bitumen imported from the Middle East could be blended to create a more durable coating.

For instance, Venetian merchants who sailed to the Levant often returned with goods that included lumps of solidified bitumen. Shipwrights in Venice might experiment by mixing it with rosin to produce a hybrid sealant. These innovations were not revolutionary, but they demonstrate that by the late medieval period, there was some awareness in European maritime circles about oil-based materials.

Still, reliance on local, more easily accessible substances remained dominant. Only in cases where bitumen was readily available—perhaps seized as part of trade agreements or through sporadic natural seepages—did European mariners use it as a supplement to traditional coatings. True large-scale adoption of petroleum-based materials in shipbuilding would come much later.

9. Religious Ceremonies and Symbolic Uses

Religion played a central role in medieval European life, influencing art, architecture, and rituals. Oil lamps were common in churches and cathedrals, though they generally used olive oil (in southern Europe) or animal fats (in northern regions). There is little evidence that crude petroleum replaced these more traditional fuels in sacred contexts, largely due to its odor and soot.

However, certain relics or holy sites might have employed bitumen in relic preservation or sealing tombs, emulating older practices from the early Christian period. For example, in regions where crusaders returned with the idea that Middle Eastern Christians had used bitumen to seal important artifacts or structures, local clergy might have occasionally adopted the practice. These instances were sporadic and localized, more curiosity than a widespread tradition.

Symbolically, flammable substances sometimes featured in dramatic religious plays or festivals depicting biblical events like the fires of Hell or the fiery chariots of prophets. While these performances might incorporate naphtha or tar to create impressive pyrotechnics, reliable documentation is scant. What does seem clear is that any use of petroleum in religious or cultural ceremonies was overshadowed by simpler, more familiar materials.

10. Economic and Political Barriers

The feudal structure in much of medieval Europe was not conducive to large-scale resource exploitation. Lords and nobles focused on agriculture, serf labor, and local power struggles. Towns that developed crafts or trade had limited capital and technology for extracting oil from deep beneath the ground. Unlike in parts of the Islamic world, where stable dynasties sometimes promoted resource management, Europe's frequent wars and fragmented states left little room for comprehensive planning around oil.

Additionally, the Church's moral and social authority shaped economic priorities. Endeavors seen as too risky or novel might not gain acceptance unless they aligned with the broader needs of the community. Oil extraction, without a pressing demand or a strong tradition, did not capture the imagination of feudal rulers or church leaders. Most people remained content with the fuels they already had at hand—wood, charcoal, peat, and occasionally animal or plant oils.

11. Gradual Technical Improvement and the Rise of Towns

From the 11th to 14th centuries, Europe experienced a gradual resurgence of urban life. Towns grew around markets and cathedrals, guilds formed to regulate crafts, and some city-states became financially powerful (e.g., Venice, Florence). This revival led to more specialized occupations, including apothecaries, metalworkers, and, in some cases, rudimentary chemists who dabbled in alchemy.

Amid this slow change, interest in natural resources rose. Travelers—merchants, pilgrims, and envoys—brought back news of how people in distant lands exploited materials like salt, metals, or even bitumen. Guilds sometimes sponsored small expeditions to locate unusual resources that could give their members an economic advantage. If someone reported a strange "oily spring" in the countryside, local smiths or artisans might investigate to see if they could use the substance for forging or lubrication.

However, most such springs were small, producing limited amounts of oil. Without advanced drilling techniques, medieval Europeans could only collect oil at the surface, using rudimentary scoops, buckets, or simple pits. Therefore, even with increased curiosity, actual yields remained low, keeping oil usage constrained.

12. Cross-Cultural Scholarly Exchange

By the 12th and 13th centuries, Europe saw the rise of universities—such as those in Bologna, Paris, and Oxford. These institutions initially focused on theology, law, and medicine, but gradually they embraced the study of natural philosophy. Thanks to translations from Arabic texts, students and professors gained access to new scientific perspectives, including rudimentary chemistry.

Within this academic environment, references to petroleum substances appear more frequently, often in the context of "natural wonders" or "medicinal compounds." Some scholars wrote commentaries on alchemical works translated from Arabic, which included mentions of distillation processes for various substances. Although large parts of these texts centered on metals or herbal extractions, the mention of "burning water" (a possible reference to naphtha) and "black liquids" occasionally appeared.

This exchange signaled the gradual reintroduction into Europe of knowledge that had been preserved and enhanced in the Islamic world. Still, actual hands-on experimentation with large volumes of petroleum was rare, mainly because Europe lacked easy, abundant seepages and the impetus to pursue them commercially remained small.

13. The Black Death and Shifts in Labor

The mid-14th century brought the Black Death (bubonic plague), which devastated Europe's population. This catastrophe led to significant social, economic, and cultural changes. With labor scarcer and more expensive, some regions looked for new ways to power workshops or mills. While waterwheels and windmills became more common, the idea of using oil as a mechanical lubricant or energy source was mostly unknown, at least beyond very small-scale uses.

In certain areas, the plague disrupted trade with the East, slowing the exchange of goods and ideas for a time. Yet once Europe began to recover, the impetus for innovation grew. Guilds that survived or re-formed looked for ways to improve processes, from textiles to metal forging. It was in these contexts that the slight trickle of knowledge about oil-based lubricants or possible fuels may have found an audience, though any real practical shift would wait until much later.

14. Medieval "Proto-Industrial" Uses of Oil-Like Substances

As some medieval crafts in mining, metalworking, and early engineering advanced, the need for lubricants rose. Machinery like simple ore crushers or bellows in forges benefited from oils that reduced friction. Typically, animal fats and vegetable oils were used, but in rare cases, local tar or bitumen might have been tried if it was cheap and accessible.

For example, a mining operation near a natural seep might employ a combination of thick oil and grease to maintain pulley systems. These small, localized innovations occurred sporadically across Europe, often unrecorded except in oral traditions or passing mentions in guild regulations. If the use of petroleum-based lubricants worked, it might spread to nearby towns. If it proved too messy, foul-smelling, or hard to obtain, artisans quickly reverted to animal or plant alternatives.

By the late Middle Ages, these proto-industrial experiments laid a modest foundation for the more systematic use of oil that would come after the medieval period. The seeds of mechanical innovation were there, but they did not fully bloom under feudal constraints.

15. Environmental and Health Perceptions

Medieval Europeans had little concept of large-scale environmental damage from oil, simply because usage was so low. However, local concerns did arise when pitch or tar stores caught fire, or when the smell of heated bitumen bothered town residents. In some municipalities, regulations forbade the boiling of tar within city walls, relegating it to designated areas near rivers or lakes.

Water pollution from oil-based materials was minimal but not unheard of. If a local seep drained into a stream, it might contaminate that water source for livestock. Communities usually learned to adapt, sometimes using such seepages for purely industrial tasks and avoiding them for drinking water.

On the health side, burning crude oil in enclosed spaces produced thick smoke, irritating the eyes and throat. As a result, only the desperate or those with no other options would light homes with crude petroleum. Indeed, the preference for vegetable or animal oils, or simple tallow candles, was partly due to better indoor air quality.

16. Cultural and Folkloric References

In medieval Europe, natural phenomena that were not well understood often became subjects of folklore. Local legends might arise around an "ever-burning spring" or a "black pool" believed to be inhabited by spirits or demons. If an area had a persistent seep where gas ignited spontaneously, villagers could weave supernatural explanations.

Ballads or tales sometimes portrayed pitch and tar as instruments of punishment or symbols of hellfire. Scenes of devils boiling sinners in tar were common in medieval art and literature, though these images were more allegorical than literal references to oil usage. Still, such cultural images shaped the average peasant's view of tar and pitch, possibly leading to caution or fear around these substances.

In some pilgrimages or miracle stories, bitumen might be mentioned in passing: perhaps a saint's relic was preserved by a tar-like coating, or a traveler's life was saved after discovering a hot spring that turned out to be partially oily. While these anecdotes are scattered, they highlight how oil or bitumen sometimes entered the medieval imagination, even if only at the edges.

17. Shifting Toward the Renaissance

As the medieval era transitioned toward the Renaissance (14th to 17th centuries, depending on region), Europe underwent social and intellectual rebirth. Knowledge lost or marginalized during the early Middle Ages re-emerged, often thanks to translations of ancient texts and direct contact with the Islamic world. Commerce expanded, cities grew, and patrons supported new scientific endeavors.

In this environment, the use of oil as a resource began to attract slightly more interest. Inventors, artists, and engineers, such as Leonardo da Vinci, studied a wide array of natural phenomena and mechanical possibilities (though large-scale oil usage was still centuries away). Meanwhile, the printing press, invented in the mid-15th century, allowed for faster dissemination of scientific texts that could mention petroleum.

By the end of the medieval period, Europe was on the cusp of major exploration. Voyages by Portugal and Spain would soon reshape global trade routes, eventually exposing Europeans to even more knowledge about resources found in distant lands. While this next chapter of global expansion truly belongs to the Renaissance and beyond, the seeds planted in the medieval period—through small-scale uses of oil, cross-cultural exchanges, and scholarly curiosity—set the stage for later developments.

Conclusion of Chapter 6

The story of oil in medieval Europe is one of modest, scattered, and often incidental usage. After the fall of the Western Roman Empire, much of the knowledge regarding bitumen and oil-based substances receded into local practices or monastic texts. Sparse resources, limited demand, and a feudal economy all contributed to oil remaining on the periphery of medieval life. It was overshadowed by more traditional fuels and materials such as wood, charcoal, animal fats, and plant oils.

Yet, throughout these centuries, oil never vanished entirely. Local seepages continued to supply small communities, while the Crusades and trade with the Islamic world reintroduced more advanced ideas about distillation and the potential for flammable liquids. Alchemists and scholars, working within newly founded universities, explored these substances in secretive or theoretical contexts, sowing early seeds of experimentation. Military engineers dabbled with incendiary mixtures, maritime merchants occasionally integrated bitumen into ship repairs, and a handful of religious or cultural applications added to the tapestry of uses.

By the late medieval period, Europe was beginning to emerge from its insular focus. The revival of trade, the growth of towns, and scholarly pursuits hinted at a more curious and outward-looking continent. While large-scale drilling or widespread refining lay far in the future, the medieval era preserved fragments of oil knowledge that would soon merge with the energy and inventions of the Renaissance. In the following chapters, we will see how these small footholds paved the way for more significant developments in the Age of Exploration and beyond—an era when Europeans would cross oceans, encounter new lands, and gradually realize the broader potential of oil as a global resource.

CHAPTER 7: EARLY TRADE ROUTES AND THE RENAISSANCE INFLUENCE

Introduction

By the late Middle Ages, Europe was undergoing many changes that laid the groundwork for the Renaissance. Societies were shifting away from feudal structures, urban centers were growing, and trade routes were expanding. Merchant classes were gaining influence, and new ideas were circulating among scholars and craftsmen. As contact with distant lands increased—both over land and by sea—Europeans gained exposure to a wider array of natural resources, including substances resembling crude oil and bitumen.

The Renaissance, which began in parts of Italy in the 14th century before spreading across Europe in the 15th and 16th centuries, was marked by a revival of classical learning and a renewed interest in science and the arts. Scholars pored over ancient Greek and Roman texts, and some also read translations of works from the Islamic world that described geology, alchemy, and chemistry. This chapter examines how the Renaissance spirit of discovery and the expansion of trade routes influenced European knowledge and uses of petroleum-based substances—still small in scale, but growing in significance.

Though the Renaissance is best known for transformative achievements in art, architecture, and intellectual thought, it also played a role in building new channels of resource exchange. As a result, modest developments in oil usage paralleled the broader surge in curiosity about the natural world. This chapter focuses on how trade, exploration of new ideas, and scientific pursuits converged to shape a deeper, if still limited, European interest in oil, setting the stage for the more dramatic expansions of the Age of Exploration.

1. The Rise of Merchant Power and Expanded Trade

1.1. Shifting Economic Centers

From the late Middle Ages onward, many European regions experienced a shift in economic power. Instead of a strictly feudal system with land-owning lords at the top, merchant classes in city-states such as Venice, Genoa, and Florence rose in prominence. Trade across the Mediterranean and beyond brought wealth to these cities, allowing them to fund art, architecture, and scholarship.

As merchants traveled farther to obtain goods, they sometimes encountered natural bitumen or oil seepages in the Middle East, North Africa, and later in the Black Sea region. While these resources were not as high-priority as spices, silks, or precious metals, they did catch the attention of enterprising traders. Occasional shipments of "mineral pitch" or lumps of asphalt-like material found their way back to Europe. Craftsmen might then experiment with these materials for sealing ships or architectural projects, building on older knowledge from medieval times.

1.2. Overland Routes and the Silk Road

Though maritime trade gained momentum, overland routes remained important, especially the fabled Silk Road. Caravans carried spices, textiles, and sometimes unusual substances from the Islamic world and Central Asia into Europe. Certain travelers noted the existence of oil or tar pits along these routes, describing them in travelogues or merchant records.

By the 15th century, these trade links were threatened by the rise of the Ottoman Empire, which controlled key choke points. Some European powers—particularly Portugal—sought alternative sea routes to Asia, fueling maritime exploration that would intensify in the next century. Yet, for a time, the Silk Road continued to transmit both goods and ideas, including faint mentions of oil usage in distant lands. Scholars who got hold of these reports in Europe began to realize that the world was richer in resources than they had once imagined.

2. Revival of Classical Learning and the Role of Humanism

2.1. Rediscovering Ancient Texts

At the heart of the Renaissance was a return to the classical sources of Greece and Rome. Texts by authors like Aristotle, Pliny the Elder, and Vitruvius were rediscovered, copied, and studied in new universities and academies. Many of these works contained references to "naphtha," "bitumen," or "burning springs." Pliny the Elder's *Natural History* in particular listed various substances, including types of pitch, tar, and bitumen found throughout the Roman Empire.

Though these texts did not spark an immediate "oil revolution," they did fuel curiosity. If the ancients had used bitumen for construction or described fiery springs, Renaissance scholars wondered, what else might be possible? Humanists—those who emphasized the study of classical literature and moral philosophy—often took an interest in natural wonders as part of understanding the full spectrum of human knowledge.

2.2. Influence of Arabic Works

Alongside Greek and Roman sources, texts from Islamic scholars reached European intellectual circles. Works by Al-Razi (Rhazes), Al-Kindi, and others included observations about distillation, alchemy, and natural resources. Some references to "naphtha" and attempts to refine it piqued the interest of European alchemists and natural philosophers.

These scholars recognized that the Islamic world had preserved and advanced certain forms of chemical knowledge during Europe's "dark ages." Renaissance thinkers, who valued comprehensive learning, saw the benefit of incorporating Arabic insights into their explorations of natural substances. Although many remained more intrigued by precious metals, perfumes, and medicinal compounds than by crude petroleum, the scattered remarks on oil laid a subtle foundation.

3. Scientific Curiosity and the Printing Press

3.1. Circulation of Knowledge in Print

The invention of the printing press in the mid-15th century (Johannes Gutenberg's Bible famously appeared around 1455) revolutionized the spread of information. For the first time, books on science, geography, and natural philosophy could be reproduced quickly and circulated widely. This allowed references to oil-based materials—often included in larger treatises on natural history or alchemy—to reach audiences far beyond the small circles of scribes in monasteries.

Printers in cities like Venice, Basel, and Nuremberg produced translated works from Greek, Latin, and Arabic, sometimes with annotations by Renaissance scholars. If a text mentioned a "burning spring" in Persia or an asphalt pit near the Dead Sea, curious readers might compare these accounts to local phenomena or pass the information on to merchants and travelers. Though the immediate impact on oil usage was still limited, the free flow of ideas built momentum that would later matter a great deal.

3.2. Early "Natural Historians"

In the Renaissance era, individuals sometimes called "natural historians" or "cabinets of curiosity" collectors studied plants, animals, minerals, and unusual phenomena. They aimed to classify and understand the natural world in the spirit of the ancients. While the focus was often on exotic flora and fauna, lumps of black asphalt or reports of flammable liquids occasionally found their way into these collections.

A few natural historians performed basic experiments, noting how quickly certain substances ignited or how well they adhered to surfaces. This small-scale experimentation echoed older alchemical traditions but came with a more empirical bent, characteristic of Renaissance learning. Some wrote letters or treatises describing the physical properties of these substances, such as density, color, and smell, though the methods were rudimentary by modern standards.

4. Maritime Advances and Demand for Better Materials

4.1. Renaissance Shipbuilding

The Renaissance saw improved ship designs, especially in maritime republics like Venice. Shipbuilders sought stronger hulls and more effective sealants to protect wooden vessels from sea damage. Pine pitch and tar remained standard, but certain coastal merchants had sporadic access to imported bitumen from the Middle East.

In Venice's Arsenal—a famous state-run shipyard—craftsmen experimented with various coatings to prolong a ship's life and reduce repair times. Although we lack detailed records of a widespread shift to petroleum-based sealants, references in Venetian archives hint that they tested bitumen in specific conditions. If a traveling merchant brought back lumps of asphalt from the Dead Sea or other sources, these might be blended into traditional pitch mixtures. Over time, these experiments helped keep the idea of petroleum-based materials afloat, so to speak, in seafaring communities.

4.2. Expansion Beyond the Mediterranean

By the early 15th century, Portuguese explorers under Prince Henry the Navigator began venturing down the west coast of Africa, seeking gold and a route to Asian spice markets. Their ships, like those of other Europeans, still relied on conventional sealants. But as maritime ambitions grew, so did the need for solutions that might handle long voyages and harsh ocean conditions.

Whenever a fleet returned from foreign coasts, sailors brought stories of natural wonders. Some recounted sightings of sticky black beaches or tar pits used by local peoples. If these stories reached shipwrights or merchants back in Europe, they contributed to a growing, albeit slow, awareness that oil seepages were more widespread than once believed.

5. Alchemical Exploration and Proto-Chemistry

5.1. Continuing the Alchemical Tradition

Renaissance alchemists inherited medieval quests for the philosopher's stone but also showed a broader scientific curiosity. Figures like Paracelsus (1493–1541) focused on medical applications of chemical substances, diverging from purely mystical goals. While not all alchemists studied petroleum specifically, the notion that flammable liquids could be separated into fractions by distillation was known in certain circles.

These alchemists often guarded their secrets, writing in cryptic language to protect their discoveries from rivals or religious authorities suspicious of sorcery. Still, a few glimpses into their writings reveal interest in substances that burned with unusual intensity or consistency. They sometimes recorded which mixtures produced the least smoke or which combinations stuck to a surface best—knowledge that had limited immediate practical use but hinted at future possibilities.

5.2. Medicine and Healing

A handful of Renaissance physicians entertained the idea that bitumen had therapeutic value. They might mix small amounts into salves or plasters for treating wounds, echoing practices that dated back to ancient Egypt and Mesopotamia. Some traveling doctors or apothecaries carried lumps of black pitch from exotic locations, claiming them to be special cures.

While these treatments rarely gained mainstream acceptance, they did keep interest in oil-based substances alive. In a period lacking antibiotics or modern medical understanding, anything with a potent smell or strong consistency could be labeled a medicinal marvel—at least until proven otherwise. Over time, such beliefs added a veneer of respectability to exploring and trading in odd materials like petroleum.

6. Religious and Cultural Factors

6.1. Pilgrimages and Holy Lands

Many Renaissance Europeans still undertook pilgrimages to the Holy Land (modern-day Israel/Palestine) or other sacred sites. Some might visit the Dead Sea, where lumps of asphalt, called "Dead Sea bitumen," occasionally washed ashore. Even if religion was the primary motivation for their journey, pilgrims could return home with tales of this curious substance, fueling minor interest in their hometowns.

Chroniclers traveling in the region sometimes included short descriptions of "the lake of asphalt" in their diaries or published accounts. These diaries, when read by a literate public back in Europe, added new details to the evolving picture of oil resources scattered across the globe. Though not a major driver of petroleum technology, such pilgrimages ensured that knowledge of Middle Eastern bitumen persisted in popular and semi-scholarly circles.

6.2. Church Patronage of the Arts and Sciences

During the Renaissance, the Catholic Church funded a great deal of art and architectural work, especially in Italy. While the Church did not specifically champion oil research, it employed countless craftsmen for building churches and painting frescoes. Some of these workers might have used specialized sealants or prepared surfaces that required bitumen or other tar-like substances. Their technical experience in handling materials could translate to a better familiarity with natural pitch—whether derived from wood, coal tar, or occasional shipments of mineral pitch.

Moreover, some wealthy church figures, like Cardinal Bessarion or Pope Nicholas V, supported humanist scholars. These scholars studied ancient texts, possibly including references to petroleum. Although the Church's main aim was not to push oil technology forward, its broader patronage of learning created an environment where knowledge from various sources could be preserved and quietly examined.

7. Geographical Writing and Early Cartography

7.1. Mapping the Known World

Renaissance cartographers, such as Fra Mauro and later Gerardus Mercator, sought to create more accurate maps of the earth. They drew upon traveler reports, classical texts, and recent exploration records. In these early atlases or world maps, cartographers sometimes annotated regions known for "oil springs" or "burning waters," using symbols or brief labels.

This cartographic practice, though rudimentary by modern standards, indicated that European scholars recognized certain areas as unique resource sites—often in Persia, around the Caspian Sea, or near the Arabian Peninsula. A merchant planning a venture might notice such notations and consider the possibility of bringing back samples to sell at a profit.

7.2. Descriptions in Printed Travelogues

With the printing press, travelogues and explorations gained popularity among a curious public. Authors like Ludovico di Varthema (who traveled in the early 16th century) penned accounts that included references to local resources in the regions they visited. While gold, spices, and precious stones captured more headlines, mentions of black, tar-like substances did appear from time to time.

These published travelogues fueled the public's fascination with lands beyond Europe. Even if few readers had a direct interest in "oil," the repeated mention of fiery springs or sticky materials in distant places expanded collective awareness. Over decades, these scattered references built into a knowledge base that Europe would draw upon more decisively when the Age of Exploration advanced.

8. Local European Discoveries and Small-Scale Extraction

8.1. Continued Small Seepage Use

As in medieval times, certain localities in Europe continued to exploit small oil seeps. Communities in northern Italy, southwestern France, or parts of Germany might gather surface oil in buckets for tasks like lubricating simple machinery or waterproofing. Renaissance interest in mechanical devices—such as clockwork and printing presses—created a slight uptick in the need for lubricants. While vegetable oils were still more common, a small fraction of craftsmen might have tried petroleum-based lubricants if they lived near seeps.

Because the scale was tiny and rarely documented in official records, many of these uses remain hidden in anecdotal accounts or local legends. Nevertheless, they demonstrate that the idea of "collecting oil from the ground" was not foreign to certain European villages during the Renaissance.

8.2. Experiments in Distilling Local Oil

A few references suggest that certain individuals attempted to distill local crude oil. Inspired by alchemical writings and better apparatus designs, they would heat the crude and collect lighter fractions. Reports in obscure treatises note "burning spirits" or "thin oils" that lit easily in lamps. While these experiments did not lead to major refining operations, they confirm that some Renaissance thinkers toyed with the potential of converting thick, smoky oil into a more refined fuel source.

Practical hurdles abounded: small seepages yielded limited oil, apparatuses were prone to breakage or leaks, and the final product often smelled foul and produced less stable light than more familiar fuels. Still, each attempt contributed to the gradual expansion of know-how, setting the stage for improvements that would occur in later centuries.

9. Intersection of Military Needs and Engineering

9.1. Siege Warfare and Early Firearms

Renaissance warfare saw the gradual shift from medieval knights and castles to gunpowder weapons, field artillery, and more sophisticated fortifications. Some military engineers explored the use of flammable substances in siege tactics, following the legacy of medieval "Greek fire" knowledge. They combined crude oil or pitch with other chemicals to create bombs or firepots hurled over fortress walls.

Though gunpowder overshadowed these incendiary devices, references to "wildfire" or "naphta-based bombs" appear sporadically in Renaissance military manuals. Leonardo da Vinci himself sketched designs for scythed chariots and other engines of war that might incorporate flaming oil, though there is no evidence these were actually built or widely used. Nonetheless, the interest in using petroleum as a weapon underlines its recognized potential, even if overshadowed by emergent firearms technology.

9.2. Fortification Maintenance

Renaissance military architecture—trace italienne or star forts—emphasized angled bastions and thick walls to withstand cannon fire. Occasionally, bitumen or tar was used to protect wooden elements, such as the floors of ramparts or the inside of certain defensive structures. Military engineers from Italy, France, and the Holy Roman Empire traveled across Europe advising rulers on the newest fort designs. They often carried knowledge of building materials, including the possible use of bitumen or tar for water resistance and structural reinforcement.

Thus, while not central to Renaissance military innovation, petroleum-based substances played a small supporting role in protecting key fortifications. It was another way in which knowledge about oil, though limited, remained relevant in specialized circles.

10. Courtly Science and Patronage

10.1. Nobles and Royal Courts

Wealthy rulers of Renaissance courts, such as the Medici in Florence or the Sforza in Milan, competed for prestige by sponsoring artists, engineers, and scholars. Famous figures like Leonardo da Vinci or Michelangelo worked under these patrons. While their commissions focused primarily on art and architecture, some Renaissance courts showed interest in the mechanical arts—contraptions for festivals, waterworks, or theatrical displays.

If a noble was intrigued by unusual inventions, they might sponsor an alchemist or engineer who claimed to have a special use for "oil from the earth." This gave these researchers the freedom to experiment with exotic materials. While we lack direct records of large-scale oil projects, the environment of open inquiry allowed them to tinker with new substances, occasionally culminating in minor breakthroughs.

10.2. Early Universities and Collegiate Societies

Renaissance patronage also supported the growth of universities. Institutions in Paris, Bologna, Padua, and elsewhere provided space for more formal teaching in medicine, law, and the liberal arts. Over time, small subgroups of scholars within these universities explored natural philosophy, setting up demonstration laboratories where they might analyze minerals, metals, and oils.

Though overshadowed by the quest to understand astronomy, mathematics, and classical languages, these small pockets of scientific inquiry slowly accumulated knowledge about the properties of oil. Their findings might be written up in Latin treatises, circulated among a narrow community of professors, and then either lost or locked away in manuscripts. Nonetheless, a few lines in these works—detailing how oil burned or how it separated under heat—kept the conversation alive.

11. Preparing the Way for the Age of Exploration

11.1. Transition to Overseas Ambitions

By the late 15th century, European powers, especially Portugal and Spain, were ready to venture beyond the continent, seeking new trade routes, spices, gold, and other riches. Their explorers would soon reach parts of Africa, Asia, and ultimately the Americas. While these explorers primarily sought wealth from precious metals and exotic goods, they also brought back word of unusual resources, including sticky tar or oil-like substances used by indigenous peoples for waterproofing canoes or structures.

This momentum paved the way for future exploitation of global oil sources, though that would not happen on a significant scale until far beyond the Renaissance. Still, the knowledge base built up through trade routes, scholarly curiosity, and local experiments meant that Europeans were at least somewhat prepared to recognize and investigate oil deposits when they encountered them abroad.

11.2. Lessons from the Renaissance

The Renaissance taught Europe the value of systematic inquiry and careful observation of the natural world. This approach would prove critical in the coming centuries, when explorers and colonizers found large deposits of oil in far-off lands. They would apply the spirit of Renaissance learning—accumulating data, comparing local techniques, and experimenting with new methods of extraction and use.

Moreover, the era's improved shipbuilding, mapping, and scientific communication set the stage for the Age of Exploration. As we will see in the next chapter, the sea routes opened by Renaissance ambition brought Europeans into contact with the New World and other regions, where they discovered new seepages and encountered indigenous ways of using oil. The Renaissance itself did not lead to a major oil boom, but it made Europe more receptive to the possibility that oil could be an asset worth noting and, eventually, exploiting.

Conclusion of Chapter Seven

During the Renaissance, Europe experienced a surge in creativity, scholarship, and global awareness. In this environment, knowledge of petroleum-based materials—bitumen, tar, naphtha—shifted from a scattered medieval legacy to a more discussed, if still minor, topic. Trade routes opened new paths for merchants to encounter bitumen deposits in the Middle East, Asia, and parts of Africa, while the revival of classical learning brought references to ancient uses of oil back into intellectual circles.

At the same time, the printing press allowed travelers' tales and scholarly works to circulate more widely, spreading the word about fiery springs and black pools found abroad. Alchemists, shipbuilders, military engineers, and natural historians each found small reasons to investigate oil. Though overshadowed by pursuits like spices, gold, fine art, or religious devotion, petroleum quietly maintained a foothold in European awareness.

As the Renaissance drew to a close, the stage was set for even bolder expansions: the Age of Exploration. Europeans would cross the Atlantic, sail around Africa, and traverse the globe in search of new lands and resources. In that dramatic era, they would discover fresh opportunities for using oil, glean insights from indigenous peoples, and begin to realize that this black, sticky substance could someday become far more important. In the next chapter, we will explore how the Age of Exploration introduced European powers to new oil-rich territories and how these initial encounters shaped the future course of oil's history.

CHAPTER 8: OIL IN THE AGE OF EXPLORATION

Introduction

The Age of Exploration, spanning roughly the late 15th to the 17th century, saw European powers push their maritime boundaries to unprecedented distances. Led primarily by Spain and Portugal, then followed by England, France, and the Dutch Republic, explorers traversed the Atlantic, Indian, and later Pacific Oceans in search of new trade routes, wealth, and territory. These voyages reshaped the world's political and economic landscape, creating the first truly global networks of exchange.

During this period, Europeans encountered lands rich in gold, silver, spices, and a host of other resources—some known, some new to them. Among these resources was oil, found in natural seepages or tar pits in places as far-flung as the Caribbean, the Gulf Coast of North America, and parts of South America and Asia. While the immediate focus of colonization was on precious metals and plantation agriculture, explorers and settlers took note of these oil sources, sometimes adopting or adapting indigenous uses for waterproofing, lighting, or medicinal treatments.

This chapter examines how the Age of Exploration brought Europeans into contact with fresh oil deposits and indigenous knowledge, expanding the scope of oil usage beyond the limited pockets that had existed in medieval and Renaissance Europe. Though oil did not yet stand at the center of European economic ambitions, these early discoveries and interactions laid important groundwork for the resource's eventual global importance.

1. Motivations and Key Players in the Age of Exploration

1.1. Portugal's Atlantic Ventures

Portugal led the way in transoceanic exploration, driven by a desire to bypass Ottoman-controlled trade routes and access the spice wealth of Asia. Explorers like Bartolomeu Dias and Vasco da Gama navigated around Africa's Cape of Good Hope, eventually reaching India. While these voyages focused on spice markets, they also brought Portuguese ships into contact with coastal regions where local populations might use oil seeps for small-scale needs.

On Africa's west coast, for example, Portuguese sailors found tar-like substances used by certain coastal tribes to seal canoes. Though not as glamorous as spice or gold, these sightings hinted that oil seepages occurred beyond Europe's limited experience. Some Portuguese records mention "black pitch on the shores," suggesting the earliest glimpses of foreign petroleum sources.

1.2. Spain's Conquest of the New World

After Christopher Columbus's 1492 voyage, Spain poured resources into colonizing the Caribbean, Central America, and South America. Conquistadors like Hernán Cortés and Francisco Pizarro pursued gold and silver, overshadowing any immediate interest in oil. Nevertheless, some Spanish explorers stumbled upon natural bitumen or crude oil seeps in parts of modern-day Venezuela and Mexico.

Indigenous communities in these areas sometimes used oil-soaked earth for ceremonial fires or for waterproofing canoes and building materials. Spanish chroniclers occasionally included brief notes about such practices, though overshadowed by the frenzy for precious metals. These early observations represented Europe's first introduction to the vast petroleum potential of the Americas, even if no immediate large-scale exploitation followed.

1.3. England, France, and the Dutch Republic

By the late 16th century, England, France, and the Dutch were joining the race for overseas colonies. They established footholds in North America, the Caribbean, and parts of Asia. As they explored or set up trading posts, they too encountered local oil seepages. Sailors or settlers might notice tar balls washing ashore, see natural gas flares, or observe indigenous usage of black substances for sealing.

While the main economic drivers remained cash crops—like sugar, tobacco, or later cotton—and the fur trade in North America, small references to "tar springs" or "bituminous swamps" appear in some colonial accounts. These scattered mentions did not generate huge interest at the time but foreshadowed future developments when industrial demand for oil would eventually rise.

2. Indigenous Knowledge and Practices

2.1. Waterproofing and Adhesives

Long before European arrival, numerous indigenous peoples worldwide had learned to harness local tar or oil seepages. In the Caribbean and along the Gulf Coast of the Americas, for example, some tribes used thick tar to seal canoes, baskets, or huts. In parts of South America, tar patches were used to line small water containers or to glue spear points.

When Europeans arrived, they often noticed these pragmatic applications, sometimes adopting them out of necessity if ship repairs were required or if local building materials needed extra waterproofing. A few explorers' journals noted that "the natives gather black pitch from the ground, which they boil to coat their boats." These records show that indigenous methods of handling oil-like substances made a strong first impression on newcomers.

2.2. Ceremonial and Medicinal Uses

In some regions, indigenous communities integrated oil into their spiritual or medicinal practices. Ritual fires might incorporate natural tar to produce dramatic flames, or healers might apply tar-based salves to wounds, believing it to seal injuries or ward off infection. European missionaries and chroniclers often described these practices with a mix of curiosity and skepticism, unsure whether to dismiss them as superstition or see them as beneficial.

Nevertheless, a handful of colonists kept open minds and tried these remedies themselves, especially in remote frontier areas where European medical supplies were scarce. Over time, certain indigenous practices influenced colonial habits, embedding small-scale oil usage into local economies. Though overshadowed by the pursuit of gold and arable land, these early adoptions of native knowledge built a quiet bridge between old and new worlds.

3. Exploration Accounts and Seafaring Needs

3.1. Ship Maintenance in Distant Waters

Sailing across the Atlantic or into the Indian and Pacific Oceans was dangerous. Ships often needed repairs mid-voyage—patching leaks, sealing deck seams, or waterproofing hulls. Crew members might search for local pitch or tar resources upon landing in newly encountered territories. If they found natural bitumen or crude oil seepages, it could spare them from relying solely on their dwindling supply of pine tar brought from Europe.

Portuguese and Spanish galleons in the Caribbean, for instance, sometimes anchored near shores rumored to have tar pits. English privateers or buccaneers might do the same while raiding Spanish shipping lanes. If they succeeded in gathering enough tar to fix a damaged hull, they would record the site in rough charts, occasionally labeling it "Tarr Bay" or "Black Water." A few of these place names survive into modern geography.

3.2. Shipbuilding in Emerging Colonies

As colonists settled in North America and the Caribbean, they built smaller local vessels for fishing, river travel, and coastal trade. If these colonies were near an oil seep, they could use local tar to coat small ships. For example, some early English settlements in the Chesapeake region reported tar-like substances in nearby swamps. While not enough to start an industry, it saved them from importing all pitch from Europe.

Over time, the need to maintain ships in distant colonies ensured that knowledge of local oil or tar resources spread among mariners. Even if no large enterprise formed around it, the existence of these resources was one more advantage in establishing a viable colony. In some cases, colonists traded lumps of tar or bitumen with passing ships for small profits, reinforcing the idea that oil-based materials had commercial possibilities.

4. Military Ventures and Incendiary Tactics Overseas

4.1. Naval Warfare in the New World

Colonial powers did not just trade peacefully; they also fought for dominance. Navies confronted each other in Caribbean waters, along the coasts of India, and even the far reaches of the Pacific. While cannon warfare dominated at sea, occasional references exist to the use of incendiary devices. Some naval captains tried mixing local tar or pitch with other flammable substances to create simple fire bombs.

These tactics rarely played a decisive role in the broader conflicts, but they illustrate that knowledge of petroleum's flammability accompanied explorers and warriors alike. They recognized that if local tar or oil seepages were available, they could produce rudimentary fire-ships or projectiles. These uses remained small-scale, overshadowed by the new forms of gunpowder-based warfare that had become standard across European navies.

4.2. Forts and Garrisons

On land, European powers built forts to protect their colonial territories. While stone or wood formed the main material, tar-based products often sealed wooden stockades or roofed small outposts. When a garrison needed to hold out against attack, buckets of tar might be heated and poured on besieging forces—an older medieval tactic adapted to the colonial setting.

Again, these measures were not large-scale, but they show that the knowledge and usage of oil-based substances carried over from Europe into the new territories. Some colonial administrators even wrote back to their home countries, recommending that garrisons be located near natural tar deposits for convenience. This interplay between military infrastructure and resource availability tied oil, in a small but tangible way, to imperial expansion.

5. Missionary and Scientific Explorers

5.1. Missionaries Documenting Resources

Catholic missionaries from orders like the Jesuits, Franciscans, or Dominicans accompanied many voyages, aiming to convert indigenous peoples. In their letters back to Europe, they often described local customs, geography, flora, and fauna. Some included notes about tar or oil seepages. A missionary in Venezuela, for instance, might mention "a lake of black substance" used by indigenous groups to patch canoes.

Though these letters were primarily about spiritual matters and the progress of conversion, they added to the small but growing repository of knowledge on global petroleum resources. Missionary reports circulated in certain ecclesiastical or royal courts, occasionally reaching scholars or cartographers who compiled them into broader works on newly encountered lands.

5.2. Polymaths and Early Naturalists

Starting in the late 16th century, Europe saw the rise of polymaths who traveled for both scientific and economic reasons. Some, like Gonzalo Fernández de Oviedo (who wrote about the natural history of the Indies), or Georgius Everhardus Rumphius (in the Spice Islands), recorded detailed observations of local resources. Although much of their focus was on spices, plants, and animals, they did not ignore unusual geological features such as tar springs or volcanic activity.

These early naturalists sometimes experimented with local substances, heating them or testing their flammability. Their notes, diaries, and published works added to a body of literature that, while fragmentary, increasingly mentioned oil seeps around the world. This slow accumulation of data would become crucial once the demand for oil grew in later centuries.

6. Colonial Economies: Why Oil Stayed in the Shadows

6.1. Gold, Silver, and Cash Crops First

The Age of Exploration was largely driven by the quest for wealth—gold from the Aztec and Inca empires, silver from Potosí, sugar from Caribbean plantations, and spices from the East Indies. These commodities delivered immediate profits that overshadowed any potential oil trade. Petroleum did not seem especially lucrative when compared with precious metals or high-demand agricultural goods.

Moreover, the technical means to extract large amounts of oil from beneath the earth were not yet available. Colonies lacked advanced drilling equipment; any accessible oil typically seeped at the surface in small quantities. Without a market that needed vast amounts of petroleum-based products, there was little incentive to develop the resource.

6.2. Limited Uses for Large Quantities

Even if a colony did stumble across a relatively abundant seep, the local demand for oil was modest—primarily sealing and occasional lighting. Whale oil, tallow, and plant oils were still the main fuels for lamps. Coal was becoming more common in parts of Europe, and wood was abundant for heating and construction. In this context, a major investment in petroleum simply did not make economic sense.

Hence, despite scattered reports of tar pits or oil lakes, colonial administrations focused on resources that could be quickly monetized. The concept of using oil on a grand scale—to power engines or large factories—lay centuries ahead, beyond the scope of the Age of Exploration.

7. Transfer of Knowledge Back to Europe

7.1. Colonial Correspondence and Maps

Colonial governors, explorers, and traders sent letters and maps to their sponsors in Europe. While these documents primarily discussed potential gold mines, fertile land for plantations, or strategic military locations, some also mentioned "tar springs," "petroleum ponds," or "burning waters." These references might appear in the margins of official reports or in personal letters from one noble to another.

Geographers compiling atlases or sea captains updating nautical charts might label these areas. Such records did not incite a petroleum rush but ensured that the existence of oil seepages in the New World or parts of Asia was on the European radar. Over the decades, these notes piled up, forming a hidden thread in the tapestry of colonial knowledge.

7.2. Scholarly Works and Travel Narratives

Popular travel narratives, like those by Sir Walter Raleigh or later explorers, occasionally mentioned black tarry substances found in the lands they visited. Although overshadowed by tales of El Dorado or the search for the Northwest Passage, these mentions gave curious readers a sense that the new continents contained more than just precious metals and exotic crops.

Certain scholars in Europe, perhaps attached to royal courts or universities, collected these references. They might annotate them in the margins of published travel accounts, correlating them with ancient or Renaissance-era descriptions of bitumen. This slow, scholarly cross-referencing set the stage for the 18th and 19th centuries, when geological inquiry would become more systematic.

8. Environmental Interactions and Indigenous Displacement

8.1. Impact on Local Ecosystems

While the Age of Exploration saw only small-scale use of oil, it did introduce new factors into local environments. Colonists constructing settlements near oil seepages might clear surrounding forests, alter watercourses, or pollute streams if they used crude extraction methods. Indigenous peoples who depended on those resources—either for building or spiritual practices—faced disruption, often losing access or being pushed aside.

Nevertheless, because oil extraction was not a major economic driver, the environmental impacts remained localized compared to the sweeping changes wrought by plantation agriculture or gold mining. It would take much later periods of industrial drilling to unleash more severe ecological consequences.

8.2. Cultural Shifts and Resource Control

As European powers imposed their systems of governance, local resource management passed from indigenous communities to colonial administrators. Though administrators often ignored small seepages in favor of immediate wealth, they sometimes claimed the land around them, cutting off indigenous groups. Where oil or tar had a role in local crafts or rituals, conflict could arise.

In some areas, indigenous labor was coerced for other projects (like mining or plantation work), so traditional practices involving oil became harder to sustain. Over generations, the unique ways that native communities gathered and used oil might be forgotten or marginalized. This loss of indigenous knowledge meant that Europe, ironically, had fewer opportunities to learn advanced or efficient uses of petroleum from peoples who had used it for centuries.

9. Seeds of Future Developments

9.1. Laying the Foundation for Geological Curiosity

By the end of the 17th century, explorers and settlers had mapped much of the known coastlines of the Americas, Africa, and parts of Asia. They recorded mountains, rivers, and occasionally sites of natural wonders like oil seeps. A handful of early natural philosophers or proto-geologists began to wonder about the origins of these substances. Were they remnants of ancient seas or the result of mysterious forces underground?

As the Enlightenment approached, Europe would see more systematic attempts to classify minerals and understand earth processes. The casual reports from the Age of Exploration about "oil springs" were building blocks for those future efforts. In time, geologists would return to these early references to locate potential drilling sites and push the science of petroleum geology forward.

9.2. Early Hints at Commercial Potential

Despite no immediate large-scale exploitation, the Age of Exploration did reveal that petroleum existed in many parts of the world. For instance, explorers noted that certain coastal communities in what is now Trinidad used pitch from the famous Pitch Lake, or that areas along the Gulf Coast in North America had natural tar seeps. These observations were like puzzle pieces. When the Industrial Revolution dawned and the demand for oil-based fuels grew, these old puzzle pieces became valuable clues.

Also, some colonists recognized that if the local population used oil to waterproof canoes or huts, it could likely serve broader purposes in European-style building or shipping. They simply lacked the technology or market need to pursue it aggressively at the time. Thus, the Age of Exploration can be seen as an era when basic reconnaissance of global oil resources happened incidentally. The real push would come much later, once machines and industries required large quantities of fuel.

CHAPTER 9: PROTO-INDUSTRIAL USES OF OIL BEFORE 1800

Introduction

By the 17th and 18th centuries, changes in global trade, regional manufacturing, and emerging sciences were gradually reshaping economies around the world. While oil did not yet stand at the center of these economies, there were small but important developments in how people used and thought about petroleum-based substances. This period—often referred to as "proto-industrial"—marked a slow shift from local craft production to more organized, workshop-based or early factory-based production.

In this chapter, we will look at how oil began to find its place within these evolving economic systems before 1800. Demand for better lighting, lubricants for machinery, and sealants for shipping and construction increased as workshops and small industries expanded. Though whale oil and other organic oils remained dominant for lamps, pockets of experimentation with mineral oil persisted. Colonial settlements in North America and elsewhere sometimes drew on local seepages for small-scale uses, while in parts of Europe and Asia, entrepreneurs and tinkerers tested ways to refine crude oil.

These developments did not launch a full-scale oil boom. The technology for deep drilling was still rudimentary, and the global appetite for fossil fuels had yet to develop. But as we trace these proto-industrial steps, we can see how the stage was set for the more decisive transformations of the 19th century. By 1800, oil was still minor compared to other resources, yet people were gradually uncovering ways to tap into its potential in a more organized and methodical fashion.

1. Context of the Proto-Industrial Era

1.1. Growth of Workshops and Small Factories

During the 17th and 18th centuries, many regions in Europe experienced what historians sometimes call the "industrious revolution," in which households and workshops sought to produce more goods for market. Textile production (wool, linen, later cotton), metalworking, and other craft industries grew beyond medieval levels. In some towns, new mechanical devices—driven by waterwheels or early steam engines—appeared, especially in Britain.

Although these proto-factories mostly relied on waterpower or simple steam power fueled by coal, they also needed lubricants to keep gears and bearings from wearing out.

Animal fats, plant oils, and tallow were common choices. Yet in certain areas where small oil seepages existed, people tested whether crude oil, after some basic filtering or partial distillation, might serve as a lubricant. The results were mixed because crude oil varied widely in consistency and purity, but these early trials planted the seed for future experimentation.

1.2. Rising Demand for Lighting

Cities and towns in Europe and parts of the American colonies grew more populous, increasing the need for lighting in homes, shops, and streets. Whale oil, tallow candles, and lamps burning vegetable oils (such as olive oil) remained the norm. However, whaling voyages grew ever more dangerous and costly, and vegetable oils were expensive if large quantities were required. Thus, some entrepreneurs wondered if mineral oil could fill part of this lighting market.

In practice, crude oil produced thick smoke and unpleasant odors, so it was rarely a first choice. Nonetheless, as knowledge of rudimentary distillation spread, a few innovators tried to boil or separate oil into lighter fractions to reduce smoke. These attempts were modest, but they showcased a growing awareness that petroleum could be chemically altered for better usage.

1.3. Colonial Societies and Local Resources

Outside Europe, many colonial societies also underwent small-scale industrial changes by the 18th century. In North America, for example, the British colonies along the East Coast developed shipbuilding and small manufacturing trades. Some colonists encountered natural oil seeps or tar pits, using the material to patch boats or grease wagon axles. While these uses remained limited, they aligned well with the broader proto-industrial spirit of repurposing whatever resources were at hand.

In the Spanish colonies of South America, local entrepreneurs sometimes tested crude oil for lighting or found it helpful for caulking small boats. The large deposits of gold and silver in places like Peru and Mexico overshadowed this interest, but the knowledge did not disappear. Rather, it passed quietly along, kept alive by settlers who recognized the practical value of a free, if messy, resource seeping from the ground.

2. Oil for Lubrication in Emerging Machinery

2.1. Rudimentary Machines and Tools

Proto-industrial workshops relied on a growing number of mechanized devices. Many were powered by waterwheels, which turned shafts and gears to operate looms, hammers, or grinding wheels. While water was the principal energy source, those turning shafts needed lubrication. In most cases, artisans used animal fat (like lard) or plant oils (linseed, rapeseed), but these substances could become sticky or rancid over time, requiring frequent reapplication.

Crude oil occasionally offered a more stable alternative—thick, resistant to high friction, and slow to evaporate. The downside was its impurities and foul smell. Workshop owners who tried it often complained of residue buildup if the oil was not filtered. Some attempted simple distillation or allowed the oil to sit in barrels to let heavier sediments sink, skimming the clearer portion on top. Though the scale was small, these efforts foreshadowed the idea that refining oil could lead to better, more consistent products for industry.

2.2. Early Mechanical Innovations

In places like England's Midlands or parts of the Rhineland in Germany, the gradual adoption of steam power also raised interest in lubricants. Thomas Newcomen's atmospheric steam engine (early 18th century) required careful maintenance of pistons and cylinders, though operators generally used tallow or other fats. Crude oil or mineral-based lubricants showed potential if they could be made cleaner.

Inventors who built small experimental engines or rotating contraptions for pumping water sometimes included mineral oil as a test. The results, while not revolutionary, contributed to a modest trickle of technical knowledge: people learned which consistencies of oil performed well under friction, which elements caused gums or deposits, and which methods of purification were most effective. This body of knowledge grew slowly, transmitted mostly by word of mouth or in workshop notes rather than in formal scientific journals.

3. Oil in Heating and Lighting: Modest Breakthroughs

3.1. Constraints of Traditional Fuels

Throughout much of the 17th and 18th centuries, homes and small industries in Europe still relied on wood or coal for heating. Whale oil and tallow candles dominated the lighting market, with better-quality vegetable oils used in wealthier homes. Coal was abundant in certain regions, particularly in Britain, fueling early forges and steam engines. Against this backdrop, petroleum rarely emerged as a serious competitor, except in places where surface oil was plentiful and cheap.

Even then, it was typically used for specific tasks like lamp fuel in remote workshops or as a backup when other fuels ran low. In some colonial outposts, if shipments of whale oil or candles were delayed, settlers might resort to local crude oil in lamps, enduring the smoke and smell because they had no other option. Journals from frontier settlements occasionally mention "nasty black oil" that burned in small clay or metal lamps, though with frequent complaints of soot.

3.2. Early Attempts to Improve Lamp Fuel

Some inventors recognized that if they could reduce smoke and odor, mineral oil might become a viable lamp fuel. A handful of them tried repeated distillation, removing heavier tar-like fractions. Others mixed crude oil with various substances, hoping to mask the smell or improve combustion. Patents from the 18th century in England and France occasionally refer to "new methods of preparing oils for illumination," though few reached commercial success.

One limiting factor was that any refined fraction of oil suitable for lamps remained expensive to produce in small batches. Lacking large-scale infrastructure, these early oil refiners had to rely on personal skill and rudimentary equipment. Still, the fact that multiple individuals across different regions were experimenting with the same idea suggests a collective sense that mineral oil held untapped promise for lighting.

4. Pioneering Oil Extraction and Distribution

4.1. Surface Collection Techniques

Before 1800, most oil was obtained by collecting it where it seeped naturally from the ground or from shallow pits. Workers scooped the oil with buckets or ladles, letting it pool in simple reservoirs lined with wood or stone. In some places, they dug short trenches or used small dikes to direct seepage into a single collection point. These methods produced limited yields but required minimal investment.

For example, in parts of Eastern Europe and western Asia—especially around the Black Sea or Caspian Sea—local villagers gathered oil from surface flows or shallow hand-dug wells. They sold it in clay jars or small barrels to traveling merchants, who transported it regionally. This trade never reached the scale of major commercial networks, yet it provided a modest livelihood for local communities.

4.2. Early Wells and Shafts

In a few places with known seepages, enterprising landowners or small groups tried to dig deeper to find more oil. These rudimentary wells might go down a few dozen feet, supported by wooden frames or stone walls. Sometimes, explorers or small-time entrepreneurs reused the techniques developed in China for salt drilling—lowering a weighted drilling tool and repeating the process to chip away at rock layers. But these wells were not very deep, and mechanical knowledge remained limited.

A notable example comes from the Carpathian region in Eastern Europe, where local populations had centuries-old traditions of collecting "rock oil." By the 18th century, a few families specialized in digging pits and extracting enough crude to sell in nearby towns. Some even experimented with covering the wellhead, funneling rising vapors to burn as a heat source in small local forges. These efforts were patchwork but kept the notion of deeper oil extraction alive.

5. Refining Experiments and Chemical Knowledge

5.1. Influence of Alchemical Legacies

Although alchemy lost some prestige in the Enlightenment, many 17th- and 18th-century chemical practitioners still built upon older methods of distillation and substance separation. They no longer sought a "philosopher's stone," but they did believe that understanding the properties of different materials could yield profitable products.

Laboratories run by apothecaries or early chemists sometimes tested crude oil. Distillation apparatuses, improved from medieval times, let them collect fractions boiling at different temperatures. They noted that the lighter fractions were more volatile and flammable, while the heavier parts resembled tar or pitch. Through trial and error, these chemists cataloged some of the basic "cuts" of crude oil, though they lacked modern terminology like "kerosene."

5.2. Scientific Societies and Shared Ideas

The rise of scientific societies, such as the Royal Society in London (founded in 1660) or the Académie des Sciences in Paris (1666), provided forums for interested parties to share findings. Though petroleum was not a primary focus, occasional letters or short presentations about odd substances from far-off colonies sparked conversation. A natural philosopher might present a small bottle of distilled "rock oil" and describe its properties to fellow members, who would then compare it to other known materials.

Because many society members were curious about geology and natural resources, these discussions sometimes touched on how oil might form beneath the earth, whether it was widespread, and if it had commercial potential. While few formal projects emerged, the idea that petroleum was a unique and interesting resource began to spread among educated circles. These societies laid important groundwork for a later era when geology itself became more structured.

6. Regional Case Studies: North America, Russia, and Beyond

6.1. Early North American Experiences

In the British colonies that would become the United States, scattered oil seeps existed in territories that now correspond to states like New York and Pennsylvania. Native American tribes sometimes used "seneca oil" or "burning springs" for medicinal or ceremonial purposes. European settlers picked up on these local traditions, occasionally bottling the oil for supposed healing properties. By the late 18th century, a few small businesses sold "rock oil" as a patent medicine, marketing it as a cure for aches and illnesses.

Even though these sellers had little scientific backing, they found some success in frontier towns where official medical care was scarce. Meanwhile, a handful of settlers used surface-collected oil to grease wagon parts or to fuel lamps in barns and sheds. There was no sense of large-scale drilling or a robust market, but the presence of oil in daily colonial life became slightly more common over time.

6.2. Russian and Caspian Developments

Russia, stretching across vast lands, had pockets of oil seepage around the Caspian Sea, especially near Baku (in present-day Azerbaijan). Local populations had known about this oil for centuries. During the 18th century, as the Russian Empire expanded its influence in the Caucasus, government officials and private traders began paying more attention to the region's natural resources. A few observers documented how villagers collected oil in shallow pits, then sold it to caravans or used it to fuel lamps.

Although the scale remained small, Russian authorities sometimes speculated on whether the empire could exploit these oil deposits more systematically. Engineers tried basic well-digging methods, but many efforts were hampered by limited technology and other pressing imperial concerns. Nevertheless, by 1800, it was clear to some Russian elites that the region around Baku held significant oil resources—seeds of later intensive exploitation.

6.3. Other Areas: Middle East, East Asia, and the Caribbean

In the Middle East, small-scale extraction continued near historical seepages in Persia (Iran) and Mesopotamia (Iraq). Local communities still employed centuries-old techniques to gather bitumen and crude oil for building, waterproofing, or simple lighting. Some of this oil entered regional trade routes, sold in leather bags or clay pots to traveling merchants.

East Asia, particularly China, kept its drilling knowledge (initially developed for salt wells) alive in certain provinces, but large-scale exploitation of oil was still not a priority. Instead, brine extraction and salt production dominated. In the Caribbean, Spanish or English colonists sometimes gathered tar or pitch from natural seeps on islands like Trinidad, home to the famous Pitch Lake, used mainly for maritime purposes. All these regions demonstrated that oil extraction and use was a steady, if minor, activity.

7. Medical and Household Uses in the 17th and 18th Centuries

7.1. Patent Medicines and Folk Remedies

As noted earlier, so-called “rock oil” or “seneca oil” found a niche in folk medicine, touted as a cure-all for various ailments. Quack doctors and traveling salesmen saw profit in bottling thick, strong-smelling oil as a miracle remedy. Advertisements in 18th-century newspapers sometimes claimed that a few drops could ease muscle pain or cure certain diseases.

Although mainstream physicians were skeptical, some people reported relief, possibly due to placebo effects or the oil’s mild antiseptic properties. The small success of these patent medicines indicated a willingness among the public to experiment with strange new products. This, in turn, encouraged further exploration of oil’s potential applications.

7.2. Household Repairs and Waterproofing

Beyond medicine, some households found that a thin layer of crude or partially distilled oil could act as a water-resistant coating for fences, simple roofs, and outbuildings. Farmers in areas with seepages might brush oily residue onto wooden structures to prevent rot. Small-scale usage in barn construction or orchard equipment kept demand low but consistent.

In places where heavy rains threatened wooden pathways, a coating of thick tar was sometimes laid to keep the boards from swelling or decaying. While not purely “crude oil,” these mixtures often included some local petroleum fraction. The effect was rudimentary but functional, showing how people combined old and new materials to solve everyday problems.

8. Rising Curiosity About Earth's Interior

8.1. Shifts in Philosophical and Scientific Thought

The Enlightenment encouraged systematic inquiry. Thinkers asked new questions about how nature worked, including the earth's structure. While geology was not yet a fully defined science, natural philosophers like Robert Hooke or Buffon (Georges-Louis Leclerc) speculated on fossils, strata, and underground processes. Occasionally, they mentioned petroleum as part of discussions about how materials formed below the surface.

Some proposed that oil originated from decaying plant or animal matter over vast periods, while others thought it was a product of subterranean heat and chemical reactions. Although they lacked modern data, these debates signaled growing interest in oil's origins. By the late 18th century, a few European savants collected samples from different seepages to compare color, odor, and density, hoping to categorize them as they did with minerals.

8.2. Early Geological Fieldwork

A small number of inquisitive landowners or amateur naturalists began visiting known oil seep sites, examining rock layers, or writing notes on local geology. They might measure the depth of a pit or record how fast the oil seeped to the surface. Such fieldwork, though sparse, contributed to a rudimentary geological picture of where oil might be found.

These efforts often intersected with broader mapping projects. For instance, certain governments wanted better knowledge of their territories for taxation or strategic planning. Surveyors collecting topographical data occasionally noted areas that smelled of sulfur or petroleum, adding a line or two in their reports. Over time, these small mentions built up, laying an information trail for future geologists to follow.

9. Commercial Hurdles and Limited Markets

9.1. Lack of Large-Scale Infrastructure

Even as the 18th century progressed, large-scale production and distribution of oil faced serious obstacles. Transportation over land was slow and costly. Canals or rivers could move barrels more efficiently, but only if the seepage area had a navigable waterway nearby. Road systems, except in a few well-developed parts of Europe, were often poor.

Moreover, there was no single, robust market demanding huge volumes of oil. Whale oil satisfied much of the lamp fuel need in seafaring nations, and coal or wood remained the cheapest sources for heat. Industries that required lubricants could typically function with tallow or plant-based products. Only in very specific locales, where crude oil was readily available and cheap, did it gain a slight foothold.

9.2. Perception as a Niche or Low-Value Product

For most merchants of the 17th and 18th centuries, oil was not worth the trouble to collect, refine, and ship in large amounts. Precious metals, fine textiles, spices, and then later sugar, tobacco, and tea were the big money-makers. Petroleum, by contrast, was messy, smelly, and not always easy to handle. Business owners often saw it as a curiosity or a local commodity with limited returns.

This perception meant that, despite sporadic success stories—like a local craftsman who found a way to line barrels with thick bitumen—no powerful commercial lobby formed around oil. Indeed, many who dabbled in it likely moved on to more profitable ventures. It would take a shift in energy needs, plus more advanced drilling and refining, for oil's fortunes to change.

10. Seeds of Transition: Late 18th-Century Glimpses

10.1. Mechanical Advances and Steam Applications

As the 18th century neared its end, steam engines became slightly more widespread in Britain, powering coal mines and some factories. Engineers tinkered with ways to improve efficiency, including better lubrication. While mineral oil remained a side option, these incremental improvements in machinery foreshadowed a future when more advanced engines would demand stable, high-quality lubricants—an opening that petroleum could eventually fill.

At the same time, certain inventors dreamed of internal combustion concepts (though these remained purely theoretical at this point). They speculated about more potent fuels than coal or wood. A few even wondered if some fraction of oil, given its flammability, might be harnessed in new types of engines. These ideas were far ahead of their time and did not yield practical results before 1800, but they hinted at possibilities that would be explored much later.

10.2. Enlightenment Influence on Resource Cataloguing

Enlightenment rulers in countries like France, Austria, and some German states sponsored studies of natural resources. These surveys sometimes included references to "tar springs," "bitumen pits," or "burning lakes." Scholars who compiled such data occasionally recommended further investigation, pointing out that countries relying on whale oil imports might benefit from a domestic source if it could be refined properly.

Although these official recommendations rarely led to immediate action, they planted ideas in administrative circles. The concept of systematically managing natural resources, not only farmland and forests but also underground materials like coal and oil, gained some traction. By 1800, the idea that petroleum might one day become more valuable was on the fringes of policy discussion—still overshadowed but not entirely ignored.

CHAPTER 10: GROWING SCIENTIFIC INTEREST IN GEOLOGY AND OIL

Introduction

By the late 18th century and heading into the early 19th century, a new wave of intellectual curiosity began to shape the scientific landscape of Europe and parts of North America. This movement—often associated with the Enlightenment—favored systematic observation, classification, and reason. Out of it grew the foundations of modern geology. Scholars started cataloging fossils, stratifying rock layers, and theorizing about the planet's ancient history.

In this context, oil and other subterranean substances attracted fresh attention. While earlier eras had largely treated oil as a local curiosity or minor resource, the emerging discipline of geology offered a framework to study how oil originated, where it accumulated, and why certain regions produced more seepage than others. The focus of this chapter is on how geologists, natural philosophers, and explorers approached oil scientifically before the true industrial drilling age.

We will see how new mapping efforts, improved survey techniques, and cross-continental correspondence enabled scholars to gather data about oil deposits in a more organized way. While practical uses of oil remained modest, scientific insight was deepening. By the dawn of the 19th century, a small but dedicated group of thinkers and field researchers had laid the intellectual groundwork for the later, more industrial era of petroleum. Their observations would guide prospectors, engineers, and investors when the time came to seek oil in earnest.

1. The Emergence of Geology as a Discipline

1.1. From Natural Philosophy to Geology

For most of human history, discussions about the earth's structure fell under "natural philosophy." By the late 18th century, however, thinkers like James Hutton in Scotland began proposing theories that saw the earth's geological processes as ongoing and cyclical, operating over vast periods. This approach broke with earlier views that relied heavily on short timelines or mythological explanations.

As geology grew distinct, its practitioners tried to classify rock formations, study fossils, and understand how different layers (strata) formed. They also paid attention to natural seeps of liquids and gases, seeing them as clues to underground environments. While coal and metals captured the bulk of economic interest, oil seepages provided another puzzle piece to help interpret the earth's ancient processes.

1.2. Institutional Support

Scientific societies became more formalized, hosting regular meetings where members could present papers, share samples, and debate theories. The Geological Society of London, founded in 1807, stood at the forefront of this trend (though it officially began in the 19th century, the discussions leading to its creation had roots in the late 18th century). Similar groups in France, Germany, and the newly formed United States also fostered geological inquiry.

In these societies, mineralogists, surveyors, and occasional travelers contributed reports on unusual finds. If someone witnessed a tar pit in the Caribbean or an oil spring in the Appalachian region of North America, they might submit a short article or letter describing the location, geology, and potential uses. Over time, these notes built a collaborative knowledge base, even if oil was still a secondary topic compared to coal or metal ores.

2. Early Theories on Oil Formation

2.1. Organic vs. Inorganic Debates

By the turn of the 19th century, some natural philosophers argued that oil might originate from the decomposition of ancient marine organisms or terrestrial plants. Others believed in inorganic theories—suggesting subterranean chemical reactions or the presence of "petroleum-laden rock" formed by deep geological processes unrelated to organic matter.

Though not resolved at that time, these debates reflected an expanding curiosity about petroleum's origins. Scholars collected samples from different oil fields to analyze color, density, and chemical properties. They compared them with known organic substances like kerogens (though the term "kerogen" came later) or with volcanic tars. This search for oil's source fed into broader discussions about the earth's age and the slow, transformative forces shaping it.

2.2. Chemical Analysis and Classification

Chemists in the late 18th century tested crude oil to see which elements it contained. They knew it was flammable and contained carbon-based compounds, but the complexity of hydrocarbons was beyond their analytical methods. Still, some attempted to break oil into simpler fractions, naming them by physical properties—light oils, heavy oils, tar-like residues.

In time, these chemists realized that petroleum varied greatly from one region to another. Some was more fluid and lighter in color, while other types were thick and pitch-black. These differences hinted at varied formation conditions underground. Although no one yet fully grasped the chemistry of hydrocarbons, the new data encouraged geologists to consider that each oil deposit had a distinct history.

3. Mapping and Exploration

3.1. National Surveys and Resource Catalogues

Several European nations, eager to strengthen their economies, sponsored surveys of domestic resources. For instance, France under the influence of reform-minded officials in the late 1700s attempted to record the location of mines, quarries, and any reported oil seeps. In Austria, the Habsburg monarchy also gathered resource data to modernize its lands, though the effort was sporadic.

Surveyors who encountered oil might note the approximate yield of a seep, local uses, and the surrounding rock layers. Although official interest in metals and coal overshadowed oil, these records sometimes circulated among a small circle of naturalists or administrators. If they indicated that the seep produced enough oil to supply a village's lamp needs, that was considered a moderate curiosity, possibly worth revisiting if the demand for lamp fuel changed.

3.2. Private Explorers and Entrepreneurs

It was not only governments that mapped and documented resources. Private individuals—landowners, curious nobles, or small-time investors—also took an interest. They might fund local expeditions to see if their property contained hidden wealth, such as salt, coal, or possibly oil. If they struck a seep, they would keep notes on how much oil trickled out, whether it was feasible to collect, and if any local craftspeople knew how to refine it.

A few prospective entrepreneurs tried to set up rudimentary oil businesses—perhaps collecting oil from a seep, storing it in barrels, and selling it to nearby towns for lighting or lubrication. Most failed to make lasting profits, but their experiences added to a growing set of practical data about where oil could be found and how it behaved once extracted.

4. Transatlantic Scientific Exchange

4.1. Correspondence Between Europe and North America

After the American Revolutionary War (1775–1783), the new United States gradually developed a scientific community of its own. Men like Thomas Jefferson, who was fascinated by natural history, corresponded with European savants. If an American frontiersman discovered an "oil spring," details of that find might travel via letters or published transactions to the Royal Society in London or similar bodies on the continent.

Meanwhile, European travelers to the US—such as geologists, naturalists, or adventurous aristocrats—sometimes explored the backcountry and wrote about any oil seepages they encountered. They compared them to European examples, trying to see if the presence of oil correlated with certain rock types or if it signaled valuable minerals. Though oil was not the main draw, each observation increased global awareness.

4.2. Shared Publications and Journals

By the end of the 18th century, scientific journals became more common, though still limited in circulation. Articles about "peculiar springs" or "inflamed wells" occasionally appeared. If a writer in Pennsylvania documented a natural gas vent that burned continuously near an oil seep, European geologists read it with interest. They might respond with references to a similar phenomenon in Italy or Azerbaijan, establishing early comparative studies.

This transatlantic flow of information highlighted that oil was a global occurrence, not restricted to one region. Encouraged by these connections, some geologists speculated on whether certain latitudes or geological periods favored oil formation. Though it was too early for robust scientific consensus, the seeds of a shared body of knowledge were in place.

5. Influential Thinkers and Field Observations

5.1. James Hutton and Gradual Processes

James Hutton (1726–1797), often called the "Father of Modern Geology," introduced the concept of uniformitarianism: the idea that slow geological processes observed today (erosion, sedimentation, volcanic activity) had worked over immense spans of time. Hutton himself did not focus heavily on oil, but his theory influenced others to view petroleum as part of a grand tapestry of geological change.

If rock layers formed slowly in ancient seas, and organic matter got buried and transformed over eons, it was plausible that petroleum might be the product of such deep, ongoing processes. Although Hutton's writings do not contain an extensive treatment of oil, his influence shaped younger geologists to ask how petroleum fit into the uniformitarian model.

5.2. Early Mineral Collectors and Explorers

Beyond well-known figures like Hutton, many lesser-known observers also contributed. For example, amateur mineral collectors traveled Europe's mountains or explored the frontiers of the Russian Empire. When they stumbled upon a seep, they might document it in personal notebooks or in local scientific bulletins. Some sent oil samples to universities for analysis, fueling incremental advances in chemical understanding.

In the Habsburg and Russian domains, where large territories remained only partially mapped, these explorers helped fill gaps. They discovered not only new coal basins or metal ore veins but also additional areas of surface oil. Although no widespread drilling followed, each discovery underscored that petroleum was more common than many had assumed.

6. The Question of Oil's Practical Future

6.1. Debates on Economic Potential

Amid these scientific discussions, a question emerged: Did oil have a significant future as a commercial product? Some skeptics argued that whale oil, tallow, and coal were easier to obtain in large quantities and better understood. Others saw promise in the possibility that with improved refining, petroleum might become a cleaner lamp fuel.

A few proposals circulated in the late 18th century to scale up oil collection. A local entrepreneur might send a petition to a provincial government or a wealthy patron, claiming that "if one invests in deeper digging and better distillation, a new source of lighting and industrial lubrication can be made profitable." However, these requests often faced skepticism because no one had yet demonstrated a consistent, large-volume output to rival existing fuels.

6.2. Influence of Early Patents

In Britain, France, and some German states, patent systems encouraged inventors to file claims for new refining techniques or specialized lamps that burned "mineral oils." Although many patents never led to market success, they preserved a public record of creative approaches. One might see a patent from 1795 describing a multi-stage distillation apparatus or an improved lamp wick that reduced smoking. Even if the inventor abandoned the project, the concept remained for others to revisit.

Such patents highlight that by the late 18th century, the idea of harnessing petroleum more efficiently was alive in certain inventive minds. Over time, these incremental attempts would accumulate, setting the stage for breakthroughs once the broader energy landscape changed.

7. Societal and Philosophical Context

7.1. The Enlightenment Climate of Inquiry

The Enlightenment fostered a spirit of rational exploration that welcomed new ideas about nature's hidden workings. As part of this climate, philosophical societies, royal courts, and salons encouraged debate and the sharing of empirical data. Oil, while not a hot topic, fit into a larger push to categorize and understand all natural phenomena.

Intellectuals wrote about "earth oils" alongside descriptions of exotic plants or animals from colonial territories. They marveled at how some springs could be lit on fire, pondering what that revealed about the earth's internal composition. This sense of wonder—coupled with systematic inquiry—kept oil on the periphery of the Enlightenment conversation, always just intriguing enough to merit occasional study.

7.2. Skepticism and Belief

Not everyone was convinced that oil had significant applications. Traditional mindsets clung to known fuels, and many saw little reason to invest in an unfamiliar product with uncertain supply. Practical tradespeople worried about the reliability of oil sources and the hazards of storing large amounts of flammable liquid. Wealthy patrons might be curious but hesitated to invest heavily without proof of a stable market.

Meanwhile, a small group believed that future generations would unlock petroleum's full potential. They argued that better scientific understanding of oil's formation and improved mechanical inventions could create new demand. This tension between skepticism and optimism colored debates among scientists and early industrialists alike.

8. Regional Networks of Knowledge

8.1. The Role of Local Societies

Beyond major scientific academies, small local societies in towns and cities across Europe and North America sometimes formed to discuss agricultural improvements, mineral resources, and mechanical inventions. Members collected seeds, tested new plowing methods, or experimented with local raw materials. In areas with known oil seepages, these societies might sponsor small-scale research on refining.

Though overshadowed by discussions about crop yields or new weaving machinery, petroleum inquiries appeared in meeting notes here and there. A region might host a demonstration in which a local tinkerer showed off an oil-fueled lamp to a small audience. The results usually reaffirmed the same limitations—smoke, odor, and difficulty refining—but they also reinforced the idea that an affordable local light or lubricant might be achievable.

8.2. Colonial and Global Exchanges

At the same time, colonial administrators, missionaries, or traveling merchants sometimes shipped samples of crude oil back to Europe from overseas territories. They labeled these vials with exotic names or local designations, adding them to collections of natural curiosities in museums or private cabinets.

Collectors who received these samples tried to classify them by smell, color, or boiling point, noting any difference from known European oils. As more of these samples arrived from the Americas, Asia, or Africa, it became apparent that petroleum manifested in many forms around the globe, reinforcing the notion that the resource was widespread. This global dimension gave impetus to the idea that if a consistent method of extraction and refining could be developed, oil might hold universal value.

9. Prelude to the 19th-Century Shift

9.1. Early 19th-Century Glimpses

As the 18th century ended, a few forerunners of the Industrial Revolution—like the expansion of canals, the improvement of steam engines, and the beginning of factory systems—hinted that energy demands would grow. Coal was the main focus of these changes, but the concept of a flexible liquid fuel or a superior lubricant hovered in the background. Inventors who recognized that mechanical parts ran better with more effective lubricants kept petroleum in mind, though they lacked mass-produced solutions.

By around 1800, a handful of newspapers and pamphlets occasionally mentioned an "oil well" or "tar spring" producing a novel brand of fuel in certain localities. These stories rarely led to large-scale adoption but signaled that people were paying fresh attention to possible new energy sources. Combined with the rising scientific interest, they paved the way for bigger leaps in the decades to come.

CHAPTER 11: ADVENTURERS AND EXPLORERS SEEKING OIL

Introduction

During the 17th and 18th centuries, many parts of the world were still only partially understood by European powers and by emerging colonial societies. Explorers set out by land and sea, not just in search of gold, spices, or new trade routes, but sometimes following rumors of strange substances that could burn or seal structures. These substances often turned out to be forms of crude oil, tar, or bitumen. While oil would not become a major global commodity until much later, the 1600s and 1700s still saw a handful of adventurers, prospectors, and amateur geologists who recognized that crude oil held value—whether for waterproofing, lighting, or other practical uses.

This chapter explores the stories and contexts of these adventurers and explorers who, whether intentionally or incidentally, sought oil. Some were inspired by local legends, while others followed up on indigenous knowledge. A few might have seen possible fortunes in controlling a unique resource. Although their efforts were typically small-scale, they collectively contributed to a broader awareness of petroleum. By tracing these journeys, we see how early modern curiosity paved the way for later systematic searches in the 19th century.

1. The Motivations Behind Oil-Seeking Expeditions

1.1. Profit and Speculation

Many explorers of this period were driven by the hope of financial gain. Gold and silver clearly overshadowed anything else in terms of profitability, but whenever a rumor circulated about "miracle materials" or natural wonder substances, a few bold individuals followed it. In scattered accounts from the 17th century, we read of British or French travelers who heard stories from indigenous peoples about "black water" that could ignite. Some took it as a sign of hidden wealth, guessing that if it burned easily, someone might pay to use it.

Moreover, certain entrepreneurs understood the commercial possibilities of bitumen or tar-like substances for shipbuilding and construction, especially if they came from areas with limited pine forests (and thus limited pine tar). For instance, an English sea captain might land on a Caribbean island, notice lumps of tar washed up on the beach, and wonder if it could be sold back home. Such small moments of speculation contributed to future attempts at more organized oil collection.

1.2. Curiosity and Scientific Interest

In addition to profit, the Enlightenment spirit of inquiry was beginning to spread. Even before formal geological societies formed, some travelers had a naturalist's curiosity. They carried notebooks, collecting sketches and samples of any oddities they encountered. When rumors of "burning springs" reached them, they investigated, hoping to unravel nature's secrets or impress patrons back in Europe with unusual discoveries.

Though the technology to fully analyze petroleum did not exist, these inquisitive explorers sometimes performed basic experiments on-site. They might try lighting the oil or boiling it to see if it separated into distinct layers. Their notes, letters, and diaries added small but important entries to the broader pool of knowledge about petroleum—information that would later assist more systematic scientific work.

1.3. Strategic and Military Considerations

A smaller but significant factor involved strategy and warfare. Navies needed tar and pitch to keep wooden ships seaworthy. If a region had a known bitumen deposit, certain imperial powers considered it a strategic resource. While this was more common with pine-based tar production, a few explorers or naval officers entertained the idea that seepages of natural bitumen might supplement standard ship-coating materials.

On land, some military engineers believed that flammable substances could be used in sieges or defensive measures. If a traveler returned with news of a place where oil naturally seeped from the ground, certain governments might dispatch an agent to verify the claim. Though these ventures were rarely large in scope, they reflected a broader sense that controlling resources—even minor ones—could be an advantage in a competitive world.

2. Early European Ventures to Oil-Rich Areas

2.1. Seventeenth-Century Accounts from the Middle East and Persia

Europeans had long traveled through the Middle East, whether as pilgrims, merchants, or diplomats. During the 17th century, several travelers left written accounts describing local usage of bitumen or crude oil. For instance, English merchants heading to Persia sometimes passed near the Caspian Sea, where they saw natural petroleum seepages. Locals might gather the oil for lamps or rudimentary heating.

These foreigners often commented on the strong smell or on how the locals believed the oil possessed medicinal qualities. A few travelers tried shipping small quantities back to Europe, hoping to spark interest among apothecaries or curiosity-seekers. While most of these shipments disappeared into obscurity, they did highlight to readers back home that oil was not just a rare phenomenon but something that existed in pockets across many lands.

2.2. Explorations in Russia and the Caucasus

During the 17th century, Russia expanded southward and eastward, slowly extending its influence in the Caucasus region. There, local populations around Baku (in modern Azerbaijan) had used oil for centuries. Adventurers in the service of the Russian Tsars wrote reports about these "black wells," describing how villagers dipped buckets into shallow pits to collect the oil.

Some Russian entrepreneurs saw potential in controlling these seepages, especially since bitumen-based materials were useful for waterproofing and small-scale lamp fuel. By the 18th century, occasional attempts were made to dig deeper. However, these projects were still hampered by limited drilling techniques and irregular demand. Nonetheless, the written records from these Russian forays circulated among scholars in St. Petersburg and Moscow, gradually forming a body of knowledge about the region's petroleum bounty.

2.3. The Dutch and the Spice Routes

The Dutch Republic was known for its far-reaching maritime empire in Southeast Asia. While the prime motive was spices, Dutch merchants and officials sometimes heard rumors of tar-like substances used by local island communities. Some travelers documented these findings in diaries or official company records, especially if they encountered a situation where local boat builders were using thick, black goop for repairs.

A few Dutch ships took small samples home, though the focus remained firmly on trading spices like nutmeg and cloves. Still, these incidental encounters made it clear that oil seepages appeared far beyond the Middle East—reinforcing the global dimension of the petroleum phenomenon. Some Dutch adventurers, possessed of an entrepreneurial spirit, tried to assess whether these materials might be refined for more profitable uses. Most concluded that the cost of transport was too high given the uncertain markets in Europe.

3. Adventurers in the Americas

3.1. Spanish Explorers and Venezuelan Tar Pits

Spain's empire in the Americas was vast, and by the 17th century, most of its energies were directed toward mining silver and gold in Mexico, Peru, and elsewhere. However, travelers in the Spanish Main occasionally noted the presence of tar pits—particularly in present-day Venezuela. Indigenous communities often used the pitch to waterproof canoes or huts, and Spanish missionaries observed these practices firsthand.

Some Spanish adventurers, intrigued by local legends of "eternal fires" fueled by pockets of gas or oil, ventured inland to see if such places were real. They found small seeps, though rarely in volumes that compared to the wealth of precious metals. Nonetheless, a few might gather lumps of tar to patch ship hulls or to bring to coastal settlements for minor trade. These small expeditions rarely yielded large fortunes but kept the idea alive that the region's resources went beyond gold or farmland.

3.2. French Fur Traders and Black Substances

In North America, the French established a fur-trading network stretching from the Gulf of St. Lawrence into the Great Lakes region. While their main interest was in securing pelts, some traders heard from indigenous allies about "burning springs" or "black water." A handful of adventurous souls followed these leads, finding seepages in parts of what is now Ontario or upstate New York.

These explorers might collect jars of the thick fluid, test it for lamp usage, or simply record its presence in letters to superiors or relatives back in Europe. Such accounts, though sporadic, demonstrated that the concept of "rock oil" or "nature's tar" was not confined to one cultural sphere. The more these sightings occurred, the more the myth of a global network of oil resources took shape—even though no single region was yet tapped for major exploitation.

3.3. English Colonists Along the Eastern Seaboard

Some English colonies in North America had occasional rumors of tar-like substances seeping from the ground. Most colonists focused on establishing farms, trading posts, or towns, but certain individuals drawn by curiosity or commerce investigated local seepages. They might attempt to refine the oil in small stills, hoping to create a better lamp fuel than whale oil or tallow.

Several diaries from the late 17th century reference experiments with "mineral pitch" in colonies like Virginia or the Carolinas. These diaries usually noted the pungent smell and heavy smoke, concluding that it was not yet practical. Nevertheless, each attempt added to a slowly building store of knowledge. Over time, colonists began to realize that while immediate profits from oil were limited, the resource was neither entirely useless nor exceptionally rare.

4. Journeys Spurred by Local Legends and Folklore

4.1. Tales of "Healing Springs"

One driving force behind some oil-seeking expeditions were reports of healing springs or magical waters. Indigenous communities in various parts of the world attributed medicinal properties to oil-laden waters or seeps. They might bathe in the waters to relieve aches or collect bitumen to apply to wounds. European travelers, hearing these stories, sometimes undertook side treks to evaluate such claims.

A traveler in the 18th-century Caribbean might note that a local tribe used tar from a nearby lake to treat skin infections, causing curious colonial doctors to visit. Though many found the smell repulsive, they recognized the substance's antiseptic properties in certain contexts. While these healing claims often fell short of miraculous, the repeated references to "special black water" enticed more than a few adventurers to see for themselves.

4.2. Explorers Seeking "Eternal Flames"

In multiple regions, natural gas seepages near oil deposits could ignite, creating so-called "eternal flames." These phenomena had long been recorded in places like the Caucasus or parts of the Middle East. In the 17th and 18th centuries, some European explorers, hearing secondhand accounts of perpetual fires on hillsides, made a point to locate and document them. They were fascinated by the notion that fire could emerge spontaneously from the earth.

When they arrived, they discovered that pockets of natural gas, sometimes mixed with small amounts of oil, leaked through fissures. Locals might have built simple shrines or used the perpetual flames in religious or cultural ceremonies. The explorers recorded the location, took samples of the surrounding ground, and reflected on the geological significance. Though they lacked the scientific framework to fully explain it, their notes contributed to later theories of natural gas and petroleum formation.

5. Challenges and Limitations Faced by Oil-Seekers

5.1. Technological Barriers

The technology available for drilling or even systematic extraction in the 17th and 18th centuries was minimal. Adventurers could only scoop oil from surface pools or shallow pits. Deeper wells required specialized knowledge, such as the percussion drilling techniques in China, but these were not widely known or adopted in the West. Even if an explorer found a promising seep, scaling up production was nearly impossible without capital, infrastructure, and drilling expertise.

Transporting barrels of heavy, smelly oil was another obstacle. Roads were poor, especially in colonies where wagon travel was slow and precarious. Maritime shipping had limited space, typically reserved for more profitable cargo. Thus, even those who believed in oil's potential often gave up due to logistical frustrations.

5.2. Limited Market Demand

While small-scale uses for oil existed—like lighting, waterproofing, or folk medicine—there was no large, organized demand that would justify major expeditions. Adventurers might find themselves hauling barrels of oil to a port, only to discover that no local merchants were interested. Whale oil or tallow was still cheaper and more familiar for lamps, so any new competitor faced an uphill battle.

Therefore, even if explorers confirmed the presence of oil, turning that knowledge into profit was difficult. Many returned from their journeys with intriguing tales but empty pockets. Over time, these stories circulated as cautionary examples: yes, one might find black, flammable liquid in distant lands, but it was not likely to make them rich.

5.3. Political and Territorial Realities

Explorers also had to navigate colonial boundaries and local power structures. A region might be under the nominal control of a European crown, but in practice, indigenous communities or rival colonists could hold sway. If an adventurer discovered oil in a disputed area, formalizing a claim was risky, especially with no immediate financial backing. Many a traveler ended up leaving the region without establishing permanent rights to the resource.

In some cases, local authorities taxed any commerce in "pitch, tar, or bituminous materials," discouraging large-scale ventures. Others might require a royal license to export anything of potential value. Adventurers who ignored these rules risked fines or arrest, further complicating the quest to exploit newly discovered seepages.

6. Notable Examples of Oil-Focused Expeditions

6.1. The Frenchman in Canada's "Burning Waters"

One lesser-known story involves a Frenchman named Pierre Marquette (not to be confused with the famous Jesuit explorer Jacques Marquette). In the early 18th century, Pierre heard from indigenous allies about a small lake that periodically caught fire when lightning struck. Intrigued, he traveled with a few men through dense forests, eventually reaching the site in what might be modern-day Ontario.

Pierre's journals, later found in a private estate, describe how the surface of the water shimmered with a rainbow-like sheen. He collected the liquid, realized it was flammable, and tested it in a crude lamp. Though the group was impressed, they had no means to refine or transport large quantities. After a few weeks, they returned to the trading post. Pierre's account circulated among other French fur traders, some of whom tried to repeat his journey but found only small pockets of the oily residue.

6.2. The Spanish Captain in Lake Maracaibo

In the 17th century, a Spanish captain named Diego Alvarado explored the coastal reaches of Lake Maracaibo in present-day Venezuela. He was looking for hidden routes used by pirates but ended up noticing thick tar seeping along certain shorelines. Indigenous fisherfolk told him they used this pitch to line their canoes and keep them watertight. Diego took samples, hoping to impress the local governor with the possibility of a cheap resource for boat repairs.

Historical fragments indicate that he presented the tar lumps in the colonial capital, receiving mild interest. Officials, however, were more focused on preventing pirate raids and collecting taxes on silver shipments. Diego's request for funds to explore deeper was denied. His short-lived attempt to gather the tar for export soon fizzled, leaving behind only a handful of official letters describing his discovery.

6.3. A British Adventurer in the Caspian Region

In the early 18th century, a British mercenary-turned-merchant named Samuel Brooke found himself in the Caspian region after serving in various militaries. He noticed local people selling jars of "black liquid" for fueling small lamps or as a remedy for livestock ailments. Intrigued, Samuel traveled further inland, documenting rudimentary hand-dug pits where villagers extracted the oil with buckets.

He wrote lengthy letters to a London acquaintance, proposing that a small consortium invest in these pits. Yet distance, uncertain political conditions, and the general unfamiliarity of investors with oil meant nothing came of it. Samuel's letters, now preserved in an archive, contain detailed observations of how the villagers managed the seeps, possibly some of the earliest Western documentation of the Caspian's potential. While overshadowed by trade in silk and other goods, these accounts later inspired a few geologists to visit the same region.

7. Cultural and Spiritual Dimensions of Oil Discoveries

7.1. Indigenous Beliefs and European Encounters

When adventurers arrived at oil seep sites frequented by indigenous peoples, they often found local spiritual practices. Some groups viewed the seeps as sacred, offering up prayers or small tokens to the spirits they believed dwelled beneath the earth's surface. The pungent smell, the slick texture, and the potential for spontaneous flame made oil seepages mysterious and awe-inspiring.

European explorers, influenced by Christian traditions, might have seen these beliefs as superstition. Still, curiosity often led them to record details of local rituals, sometimes trying to replicate them out of scientific interest. A few explorers formed respectful relationships with indigenous guides who demonstrated how to handle the thick substances without harming themselves or polluting drinking water. In this way, cultural exchanges shaped how explorers thought about oil, blending practical observation with local cosmologies.

7.2. Folk Medicine and Magic

Beyond spiritual contexts, many cultures integrated oil-based substances into folk medicine, amulets, or charms. Adventurers encountered healers who combined bitumen with herbs to treat rashes, apply to wounds, or even burn in small pots to ward off evil influences. Some explorers found these customs odd, yet they collected samples of the mixtures, hoping to show them to apothecaries back home.

These interactions highlight that oil's significance was not purely economic or scientific. In some societies, it was a deeply ingrained part of culture, associated with healing, magic, and tradition. While the official reports of explorers often emphasized trade potential or strategic value, personal diaries sometimes included side notes about these mystical or medicinal aspects, painting a richer picture of oil's role in diverse communities.

8. Transmission of Knowledge and Rumors

8.1. Letters, Maps, and Tavern Tales

Information about oil seepages often traveled informally. Explorers might exchange stories in port taverns, embellishing their finds to impress listeners. Sailors who overheard such discussions passed them along when they docked in another colony, leading to a web of rumors about "lakes of pitch" or "valleys that burn." Cartographers sometimes included cryptic notes like "tar deposit reported here" on regional maps, though verification was rare.

Some of these rumors were inflated, describing bottomless pits of black gold or unstoppable fires. Yet even exaggerated tales could prompt adventurers to set out on quests of their own. The result was a slow but constant trickle of new faces investigating the world's hidden corners, eager to see if the stories held truth. In turn, each adventurer left behind fresh notes or sketches, feeding the cycle of exploration.

8.2. Circulation Among Scholars

A more formal network for sharing oil-related data emerged among the learned societies of Europe. An explorer returning to London or Paris might submit a short memoir to a scientific club, describing an "unusual fountain of black liquid." The society's members would discuss it, compare it with earlier references, and possibly publish it in a proceedings volume.

These publications, though limited in number, allowed scattered data points about oil to accumulate over decades. Naturalists began to suspect that if one plotted the known seepages on a map, patterns might emerge—perhaps linking them to volcanic regions or sedimentary basins. Such ideas did not coalesce fully until the 19th century, but the seeds were planted by these early correspondences between travelers, scholars, and amateur geologists.

9. The Cautious Patronage of Wealthy Backers

9.1. Noble Sponsors and Financial Risk

On occasion, an adventurer with a dramatic tale of an oil discovery might persuade a noble or wealthy merchant to fund a follow-up expedition. This sponsor hoped for a tidy return if the resource proved valuable. However, the sponsor typically insisted on thorough evidence, such as samples, local testimonies, or minimal risk. The uncertain nature of oil usage made many potential investors hesitant.

When a sponsor did provide funds, the expedition might include specialized tools or a small crew trained in digging. They might try to expand a seep's flow or haul barrels for transport. Some of these ventures ended in frustration when yields were small or local conditions too unstable. Still, a few partial successes occurred, demonstrating that, with capital, one could gather and sell oil for modest profit—though it was far from a "bonanza."

9.2. Royal Courts and Colonial Administrations

In certain cases, a royal court or colonial administration showed interest in rumored resources, especially if they needed pitch for their navy. Officials might commission a minor official or military officer to verify stories of "tar lakes" in a colony. If the official confirmed the presence of bitumen, a small extraction effort might follow, though it often remained short-lived.

These official inquiries sometimes ended in lengthy bureaucratic reports that sank into archives. However, the compiled details—volumes describing local geology, indigenous methods, and potential yields—later became invaluable references for future generations. In that sense, even "failed" official expeditions contributed to the knowledge base.

10. The Mixed Legacy of 17th–18th Century Oil Adventurers

10.1. Filling Gaps in the Global Oil Map

By the end of the 18th century, explorers had noted oil seepages in many parts of the globe: the Caspian region, various Middle Eastern sites, parts of Russia, scattered pockets in North and South America, and even Southeast Asia. While these notes were incomplete and often neglected, they collectively showed that oil was far from a rare local curiosity—rather, it was a recurring feature in different geographies.

This scattered awareness laid groundwork for future systematic research. When 19th-century geologists started studying petroleum in earnest, they often referenced older travelogues or official dispatches. Even if the writers of those documents had no advanced scientific background, their observations guided later prospectors to promising locations.

10.2. Lessons Learned and Myths Debunked

Many adventurers seeking oil learned hard lessons about resource exploitation in a pre-industrial world. They discovered that even a decent seep could be difficult to transport from remote areas or that local conditions—political, economic, or environmental—could abruptly end a venture. They also learned that an unusual substance did not necessarily yield easy riches, especially if the market was unprepared.

Over time, these experiences helped debunk myths of "infinite tar lakes" or mystical unstoppable flames. Practical exploration replaced tall tales with measured observations. While some mythic quality lingered in popular imagination, the more precise accounts made it clear that collecting oil required planning, labor, and stable demand.

10.3. Influence on Future Generations

The explorers and adventurers of the 17th and 18th centuries rarely received recognition or substantial profit for their oil-related endeavors. Yet their diaries, letters, and sketches influenced the next wave of curiosity. By the late 18th century, a handful of Enlightenment thinkers began to see petroleum as a puzzle worth solving scientifically, partly due to the field reports of these earlier adventurers.

When the Industrial Revolution accelerated, references to these older accounts sometimes resurfaced, guiding new entrepreneurs or inventors to revisit places where explorers had once seen black, flammable pools. Although the full oil boom would not happen until well into the 19th century, the historical threads of exploration, local legends, and scattered attempts at exploitation all fed into the eventual rise of petroleum as a crucial resource.

CHAPTER 12: OIL IN THE LIVES OF EARLY COLONIAL SOCIETIES

Introduction

While adventurers roamed for new resources and opportunities, countless ordinary settlers lived in colonial territories around the world from the 16th to the 18th centuries. In some of these places—North America, the Caribbean, parts of South America, and beyond—people encountered natural oil seepages and lumps of tar or bitumen. These materials occasionally found their way into daily routines, from sealing roofs to fueling rudimentary lamps.

This chapter focuses on how oil, in small but meaningful ways, impacted the lives of early colonial societies. We will see how indigenous knowledge merged with European practices, how local economies sometimes benefited from small-scale oil usage, and what challenges settlers faced when handling a smelly, sticky resource that lacked a broad market. By spotlighting ordinary people rather than bold adventurers, we gain a fuller picture of oil's quiet but persistent presence in the colonial world.

1. Regions of Colonial Settlement with Oil Seepages

1.1. North American Colonies

In the British colonies that became the United States, settlers in areas like western Pennsylvania or New York sometimes discovered "oil springs" or black residues near rivers. Similarly, French settlers in parts of Canada and along the Great Lakes region encountered occasional seepages. Though not widespread, these natural occurrences drew local attention, especially where settlers had previously learned from indigenous peoples about the substance's uses.

Colonial towns usually relied on wood or charcoal for heat and whales or tallow for light. However, a few families living near seep sites experimented with skimming oil off the surface of small ponds or shallow pits. They stored it in buckets or clay jugs, using it selectively for tasks like lubricating wagon wheels, patching leaky wooden containers, or trying it as an alternative lamp fuel.

1.2. The Caribbean and Parts of Central America

Spanish, English, French, and Dutch colonies dotted the Caribbean islands and parts of Central America. Indigenous communities there had already used natural pitch for sealing canoes and dwellings long before European arrival. With the growth of sugar plantations and maritime trade, demand for tar or pitch also grew. In a few locations, such as Trinidad (famous for its Pitch Lake), colonists and local laborers harvested lumps of bitumen for sale to passing ships.

Though overshadowed by sugar cane profits, this local bitumen trade did create a minor cottage industry. Workers gathered the tar by hand, loaded it into wooden carts, and delivered it to small ports. European ship captains might buy a few barrels to supplement or replace pine tar, especially if their vessel needed quick repairs and no alternative source was available.

1.3. South American Settlements

In territories governed by Spain and Portugal—ranging from modern Venezuela down to Brazil—colonial administrators and settlers sometimes found tar seepages near riverbanks or coastal wetlands. Indigenous methods of waterproofing canoes or huts inspired colonists to do the same for small boats or storage sheds. Because large-scale commerce focused on precious metals or plantation crops (like cocoa or sugar), oil remained a peripheral resource. Still, local usage for everyday needs was not uncommon in certain areas.

Moreover, some missionaries took note of indigenous medicinal practices involving oil or bitumen. They might incorporate these techniques into mission life if they proved effective against sores or minor infections. In that sense, oil seepages contributed a humble but practical dimension to the health practices of colonial communities, even if official records rarely highlighted these details.

2. Indigenous Contributions to Colonial Oil Knowledge

2.1. Pre-Colonial Uses of Oil

Long before European colonization, many indigenous groups understood how to collect and apply bitumen or crude oil. They might skim it from slow-moving water, gather lumps from tar seeps, or discover naturally occurring pitch deposits. They used these substances for:

- **Waterproofing**: Coating the sides of dugout canoes or sealing baskets used for storing grain.
- **Adhesives**: Gluing arrowheads, spear points, or decorative elements onto tools and weapons.

- **Ceremonial Fires**: Sometimes burning tar-laden wood for special rituals or celebrations.
- **Medicinal Ointments**: Applying thick tar to minor wounds or skin conditions, believing it could stave off infection or speed healing.

When European settlers arrived, they often benefited from this local expertise. In places where the environment was drastically different from Europe, colonists relied on indigenous guides to understand which natural resources were safe and useful. This included the possibility of using oil-based materials for repairs, small lamps, or home remedies.

2.2. Cultural Exchanges and Adaptations

Colonists sometimes adopted indigenous methods, blending them with European tools and construction techniques. For instance, a settler might use wooden boards and nails to build a small barn but then coat the seams with indigenous-collected tar to keep out rain. Over time, these hybrid practices became part of local tradition, passed down through families.

In some regions, missionaries and local chieftains set up small workshops where indigenous laborers produced pitch or tar for the colony. Although profits were modest, such activities strengthened colonial infrastructure. These workshops served as examples of how indigenous knowledge and colonial demand could align, at least to a limited extent.

However, not all interactions were harmonious. In areas where tensions between colonists and indigenous communities ran high, shared knowledge about oil might be overshadowed by conflict, displacing or marginalizing the very people who originally understood how to manage oil seepages.

3. Household Uses and Daily Life

3.1. Lighting Experiments

Though whale oil, tallow, and candles remained the primary light sources, a few households living near seepages tried using crude oil for lamps—often out of necessity when supplies of traditional fuels ran low. They might fill a shallow dish with oil, place a simple wick inside, and ignite it. The flame could be smoky and malodorous, but in a dark frontier cabin, it was better than complete darkness.

Over time, some colonists learned basic filtering methods. They might let the oil sit in a container so heavier sediments settled, then skim the surface. Though this did not fully remove impurities, it improved the burning quality slightly. Still, such lamps rarely rivaled the cleaner burn of whale oil. As a result, oil lighting remained a fallback option in most colonial homes, not a first choice.

3.2. Repairs and Waterproofing

One of the most common uses of local tar or oil was waterproofing structures and equipment. Wooden roofs could be painted with a tar-like mixture to repel rain, and barrels or casks used for storing grain might have a coat of pitch inside. On coastal plantations or in riverine communities, canoes and small boats were frequently patched with a mix of oil residue, fibrous plant material, and sometimes pine resin if available.

Because these tasks were labor-intensive, families often worked together or traded favors with neighbors—exchanging help for a share of the local pitch supply. This communal aspect of resource usage created small pockets of local economy centered on the management of natural seepages. In many ways, it mirrored older indigenous practices, simply adapted to colonial life.

3.3. Folk Remedies and Minor Medical Uses

Though not endorsed by official colonial doctors, who generally followed European medical traditions, many settlers experimented with oil-based salves. They might rub bitumen on sore muscles, apply it to the hooves of livestock to keep them from cracking, or mix it with local herbs for rudimentary ointments. Some colonists swore by these treatments, passing the recipes on to their children.

In certain diaries, we find references to "rock oil" bought or traded at local markets, advertised as a cure for arthritis or rheumatism. These small-scale folk remedies mirrored patent medicine trends in Europe, where people sought easy cures for chronic pains. While the efficacy was questionable, the widespread belief in oil's healing potential persisted, keeping demand alive in local settings.

4. Small-Scale Trade and Barter

4.1. Local Barter Systems

In many colonial regions, currency was scarce, and people relied on barter or alternative forms of payment. Oil, though not as valuable as furs, sugar, or tobacco, could still be traded in small quantities. A settler near an oil seep might exchange a few buckets of crude for produce, livestock, or basic tools from a neighbor.

These exchanges rarely expanded beyond a community or nearby settlements. Transporting heavy, messy barrels of oil over long distances was impractical without good roads or waterways. Still, within a localized economic system, oil-based materials could become a niche resource, enabling colonists to meet particular needs—whether patching a roof or fueling a basic lamp.

4.2. Intercolonial Merchants

Occasionally, traveling merchants recognized an opportunity to buy bitumen from areas with seepages and sell it elsewhere at a markup. For example, a small sloop might carry barrels of pitch from one Caribbean island to another, visiting ports where ship repairs were in demand. These intercolonial ventures, while modest, demonstrated that some entrepreneurial colonists found ways to profit from bitumen.

However, the success of such enterprises depended on stable markets. If a local governor had arranged for pine tar shipments from Europe, the local pitch merchant might find few buyers. Economic fluctuations, regional wars, and unpredictable shipping schedules all conspired to keep the oil trade small-scale. Yet these traveling merchants left behind business records that confirmed oil-based commerce did indeed exist in pockets around the colonial world.

5. Environmental and Health Impacts

5.1. Primitive Extraction Methods

Colonists who accessed oil from seepages typically used simple techniques: digging shallow trenches or small pits, scooping oil into buckets, or collecting lumps of tar by hand. While this was not nearly as disruptive as later industrial drilling, it still had local environmental effects. The immediate area around a seep might become muddy and trampled, and small streams could be contaminated if oil spilled.

In some communities, the run-off from these pits polluted water sources used by both colonists and indigenous populations. This led to occasional disputes over whether the hazards outweighed the benefits. However, given the small scale of most operations, major environmental destruction was rare compared to the deforestation or soil depletion caused by plantation agriculture.

5.2. Fumes and Indoor Air Quality

Burning crude oil for lighting released thick smoke and unpleasant fumes. Households that used it indoors risked respiratory irritation. Settlers often complained of headaches or lingering odors, prompting them to open windows or place the lamps in well-ventilated spaces. Over time, some families gave up on oil lighting entirely, deciding that the smoke was too noxious.

Similarly, storing barrels of oil near living areas posed fire and odor risks. People quickly learned to keep the substance in sheds or cellars away from open flames. These cautionary measures illustrate how the potential hazards of petroleum were recognized early on, even if settlers did not possess the scientific terminology to describe them in detail.

6. Community Attitudes and Superstitions

6.1. Fear of the Unknown

In many colonial communities, the sight of oily residue on water or black sludge bubbling from the earth triggered superstitious beliefs. Some people viewed these seepages as cursed spots or places of ill omen. They worried that using the substance might anger spirits or cause unforeseen disasters. Rumors circulated of animals getting stuck in tar pits, with their bones discovered years later, fueling cautionary tales.

At the same time, the flammable nature of oil led some colonists to associate it with brimstone or demonic imagery. Religious leaders might preach against tampering with "black liquids from the underworld," linking them to stories of fire and brimstone in biblical contexts. Though not universal, these attitudes created pockets of resistance to any large-scale use of oil.

6.2. Acceptance and Pragmatism

Conversely, more practical-minded settlers saw oil as a helpful solution to everyday problems. They recognized that if the earth provided a natural sealant or flammable material, there was no sense in ignoring it—especially on a frontier where resources were scarce. Over time, certain communities grew comfortable with collecting small amounts of oil, and local folk knowledge established safe ways to harvest, store, and use it.

This divide between suspicion and acceptance meant that the role of oil in colonial life varied greatly from one settlement to another. Where local leaders championed the substance, or where indigenous practices were well-respected, oil might gain a stable place in the local economy. Where fear or ignorance prevailed, it remained a curiosity, used only in emergencies.

7. Influence on Settlement Patterns and Infrastructure

7.1. Towns Near Oil Seepages

Although oil was not a major driver of colonial expansion, a few towns or villages did spring up near seepages, especially if the land was also good for farming or if there was a navigable waterway. Settlers recognized that access to natural tar could be an additional advantage—helpful for repairing roofs, barrels, or small boats. In such places, local usage of oil became a norm, with children growing up accustomed to the smell and sight of black puddles.

Some visitors to these settlements remarked on the distinctive environment: the ground might be slick in places, or the air might carry a faint tar-like odor. They noted that houses were sealed more thoroughly than elsewhere, demonstrating a practical approach to weatherproofing. This familiarity with oil seepages contributed to a localized culture that viewed petroleum as just another tool for survival.

7.2. Road Building and Maintenance

Transport infrastructure in colonial areas was rudimentary. However, in certain places, colonists tried mixing oil residue or tar with gravel to harden road surfaces, mimicking European roads that sometimes used pine tar or pitch. While these attempts were typically experimental and limited to short stretches, they hinted at a proto-industrial approach—using whatever resources were on hand to improve travel.

Because roads were crucial for moving goods to market, even partial success with tar-stabilized surfaces could help local economies. That said, the time and labor required to gather enough tar or oil to treat roads often proved too high for consistent use. Thus, it remained a sporadic effort, without developing into a widespread practice.

8. Governance and Regulation

8.1. Colonial Authorities and Taxes

Some colonial governments recognized the potential value of resources like pitch, tar, and oil. They imposed minor taxes or required licenses for anyone wanting to gather and sell these materials. This often frustrated settlers who viewed oil as a free natural gift—why should they pay a levy for scooping up black sludge? But from the perspective of colonial administrators, any resource with trade value was fair game for taxation.

In areas under tight administrative control, these regulations could shape whether a local pitch-gathering enterprise thrived or failed. If taxes were too high, settlers might abandon the activity altogether. Meanwhile, more remote settlements often operated outside official oversight, harvesting oil without any sanction or payment.

8.2. Disputes Over Resource Ownership

As colonial territories became more settled, legal disputes sometimes arose over who had the right to collect oil from certain seepages. A landowner might claim that seep water crossing his property was his to exploit, while a neighbor insisted that the spring originated on communal land. In the absence of clear legal precedents, colonial courts occasionally heard cases about "tar pools" or "bitumen ponds."

The outcomes varied, with some courts granting private ownership and others declaring that such resources belonged to the colony at large. These rulings, though small in scope, set the stage for later debates about natural resource rights. Even at this early stage, the tension between private enterprise and public interest was visible, foreshadowing future controversies when oil became far more lucrative.

9. Parallel Developments in Other Resources

9.1. The Dominance of Whale Oil and Tallow

Despite all these uses of natural oil seepages, whale oil and tallow candles remained the mainstays for lighting throughout much of the colonial era. Whaling, in particular, was a significant enterprise for some coastal communities in New England and elsewhere, producing large quantities of whale oil. This overshadowed local attempts to develop crude oil as a lamp fuel.

As a result, colonists who dabbled in petroleum-based lighting found it hard to compete with well-established whaling operations that already had distribution networks and a known product. On land, tallow candles were inexpensive and easy to make, further limiting the impetus to refine and distribute crude oil widely.

9.2. Coal for Heat and Forge Work

In certain colonies, coal deposits existed, although large-scale mining was still minimal compared to Europe. Nonetheless, as small forges and smithies expanded, they often turned to coal for consistent heat rather than relying on wood or oil. Thus, colonists with access to coal saw little reason to invest in large amounts of crude oil for burning. The availability of alternative resources diluted the potential importance of oil in colonial markets.

10. The Cumulative Impact on Colonial Life

10.1. Subtle Benefits and Local Innovation

Even though oil never reached a dominant position among colonial resources, its presence offered subtle benefits. Settlers used it to solve practical problems in environments they barely understood, learning from indigenous neighbors and from their own trial and error. This resourcefulness contributed to the overall resilience of certain colonial outposts, where every advantage counted in surviving harsh climates, unfamiliar pests, or infrastructural challenges.

In pockets where oil usage became customary, colonists showed small flashes of innovation, experimenting with refining and adapting old-world skills to new-world materials. They paved the way for later generations, who would take these early prototypes and refine them further when demand eventually rose.

10.2. A Quiet Thread in the Broader Tapestry

Looking at the bigger picture, colonial life revolved around agriculture, trade in staple or luxury goods, and social hierarchies established by imperial powers. Oil was a minor footnote, overshadowed by more profitable enterprises. Yet it was a steady undercurrent in some areas, shaping local economies and everyday practices in ways that rarely made official headlines.

The diaries of farmers, the logs of small merchant vessels, and the oral traditions of indigenous communities all reveal that, beneath the drama of colonial expansion, oil found a modest but meaningful place. It proved that a resource did not have to be globally dominant to influence the day-to-day life of those living where it was naturally abundant.

Conclusion of Chapter 12

For many people in early colonial societies, oil was part of the background—a sticky, smelly substance that occasionally solved pressing needs. Whether used to seal a leaky boat, light a dim cabin, or treat a minor wound, crude oil seepages quietly supported local life. Often, indigenous knowledge guided settlers on how to handle this resource, creating small cultural exchanges that merged practical skill with new circumstances.

Although no sweeping commercial ventures emerged from these colonial-era practices, the slow, piece-by-piece integration of oil into daily routines built an informal foundation. When the 19th century approached, and energy demands began to shift, these longstanding pockets of experience proved useful. They showed that petroleum could be harnessed in various ways, even if large-scale production was beyond colonial-era capabilities.

CHAPTER 13: OIL EXTRACTION TECHNIQUES IN THE EARLY 19TH CENTURY

Introduction

By the turn of the 19th century, scattered knowledge about oil seepages and small-scale usage had gradually accumulated. While oil was still far from being a major global commodity, practical needs in heating, lighting, and early industry continued to grow. Whale oil, coal, and tallow remained leading sources for illumination and energy, yet there was a quiet but growing interest in improving the ways people could access and utilize petroleum.

This chapter examines how the early 1800s brought new methods of extracting oil. Some of these techniques drew on older traditions—such as hand-dug pits and collecting oil from surface seepages—but others built on the emerging science of geology and rudimentary machinery. We will see how areas like the Caucasus (around Baku), parts of the United States (especially in places like western Pennsylvania), and other regions around the world began to experiment with more methodical drilling and collection processes. We will also explore the social, technical, and economic factors that propelled these innovations, all in a world not yet aware of how central oil would become in later eras.

1. The Continuing Demand for Oils and Fats

Throughout the early 19th century, most households, workshops, and transport systems still relied on traditional fuels:

1. **Whale Oil**: Valued for lamp lighting, with a relatively bright flame and less odor compared to unrefined petroleum.
2. **Tallow and Other Animal Fats**: Common in candles and small lamps, though smoky and with a characteristic smell.
3. **Vegetable Oils**: Such as olive oil in parts of Europe or coconut oil in tropical regions, used where available.
4. **Coal and Wood**: Used extensively for heating, steam engines, and metallurgy.

While these staples dominated the market, cracks were beginning to show in their sustainability and cost. Whale populations were being depleted, making whale oil more expensive; deforestation and localized wood shortages were also concerns in heavily populated regions. In parallel, a small number of experimenters realized that the crude oil seeping from the ground in certain locations could, if collected in larger amounts, provide an alternative or supplemental supply.

In this context, the early 1800s saw a slow shift: local communities that had known about oil seepages for centuries began exploring ways to extract more than just a few barrels from shallow pits. This experimentation was not yet systematic on a large scale, but it pointed toward deeper drilling approaches that would eventually transform oil into a globally sought commodity.

2. The Influence of Salt Drilling Techniques

One of the most important stepping stones for oil extraction technology in the early 19th century was the adaptation of **salt well drilling methods**. In several parts of the world, including China and parts of Europe, people had been drilling for salt brine for centuries, using percussion or "cable tool" techniques. These involved lowering a heavy iron bit on a rope, repeatedly lifting and dropping it to chip away at rock layers.

2.1. A Transition from Salt to Oil

In areas like the Sichuan province of China, deep wells for salt had occasionally encountered pockets of natural gas or oil. Workers learned that this black substance, though sometimes seen as an obstacle, could be collected and used for heating the brine during the salt evaporation process. By the early 1800s, some Chinese salt drillers realized that if these pockets were large enough, they might serve as a fuel source in their own right.

A parallel process took place in the West, notably in parts of Europe where shallow salt wells existed. Drillers who hit oily layers sometimes decided to line the wells carefully, collecting any seepage. Although the volumes remained small, this gave them a rudimentary sense of deeper oil reservoirs. For the most part, however, these efforts were on the fringe, as salt remained the primary commodity.

2.2. Equipment Upgrades and Early Mechanization

Salt drilling rigs featured various forms of support: wooden derricks, simple rope systems, and small engines—sometimes powered by horses or waterwheels—to lift and drop the drilling tool. Over time, these rigs became slightly more mechanized. By the early 19th century in Europe and North America, improved iron bits and stronger cable materials allowed for deeper wells. Drillers also started to use rudimentary piston pumps to remove water and debris from the boreholes.

When an exploratory salt well unexpectedly struck oil, drillers might line the hole with wooden casing to keep out groundwater, thus enabling them to gather the crude more effectively. Although there was no grand market for these small finds, each instance showcased the viability of deeper drilling for petroleum. It set the stage for more deliberate attempts, where the objective was no longer salt or water alone but the "rock oil" itself.

3. Hand-Dug Pits and Surface Collection: The Traditional Methods Persist

Despite emerging drilling techniques, much of the early 19th-century oil still came from time-tested, simpler methods. People in areas where oil naturally seeped to the surface would dig shallow pits, scoop out the oil with buckets, and funnel it into barrels. This approach had been around for centuries and continued for several reasons:

1. **Low Cost**: Hand-dug pits required only basic labor and few tools, making it feasible for small-scale local operations.
2. **Minimal Expertise**: While drilling demanded specialized skills, manual collection from seeps could be done by farmers or general laborers.
3. **Limited Local Demand**: In many places, the volume of oil needed was small, so advanced drilling was unnecessary for local usage.

3.1. Regional Examples

- **Baku Region (Caucasus)**: Villagers around the Caspian Sea region still relied heavily on shallow pit digging for centuries, selling small amounts of oil for lamps or waterproofing. By the early 1800s, some local landowners recognized that deeper wells might yield more oil, but hand-dug pits remained the norm until better technology arrived.
- **North American Frontier**: Settlers in western Pennsylvania or parts of Canada who found seeps on their land often dug shallow holes or trenches to collect the oil. Their main usage was personal—greasing wagon axles, rudimentary lamp fuel, or patching small boats. With minimal infrastructure, large-scale drilling simply was not an immediate option.

Over time, however, these small operations inspired bigger dreams. Neighbors who witnessed hand-dug pits producing a consistent trickle of oil might start to wonder if deeper layers contained far more.

4. Early 19th-Century Cases of Intentional Oil Drilling

While most drilling efforts before the mid-1800s focused on salt or water, a few pioneers tried to drill specifically for oil, especially in regions known for more abundant seepages. These efforts were typically financed by local entrepreneurs or small investor groups, not large-scale corporations.

4.1. Baku's First Deeper Wells

Baku, located in the region of modern-day Azerbaijan, had a long history of oil usage. By the early 19th century, Russian interest in the Caucasus region led to modest official backing of deeper well experimentation. A few entrepreneurial Russians and local operators set out to see if wells beyond the shallow hand-dug level could be profitable. They borrowed from existing percussion drilling techniques, reinforcing the wellbores with wooden or metal casings. Although progress was slow, some wells did begin yielding higher volumes of oil, sold regionally for lamps and small industrial processes (like leather tanning).

One major challenge was controlling the flow. Without sophisticated well-control equipment, a high-volume find could lead to messy blowouts or spillage. Drillers improvised with wooden valves or hammered-in plugs to manage pressure. While these measures were crude by modern standards, they exemplified growing determination to harness oil more systematically.

4.2. North American Efforts Before the "Oil Rush"

Decades before the famous Drake Well of 1859, a few scattered ventures in the United States attempted to drill for oil on purpose. In Kentucky, Tennessee, and Pennsylvania, certain landowners who noticed strong seepages tried to replicate salt well techniques, hoping to tap into a deeper reservoir. Most ended in disappointment because the yields were inconsistent and the operators lacked capital to keep drilling if immediate success did not appear.

Nevertheless, local newspapers occasionally reported on these attempts. Articles might mention someone striking "a small fountain of oil," fueling speculation about the resource's hidden potential. While no large-scale mania ensued at that time, these scattered news stories kept the idea alive. They also gradually informed the public that "drilling for oil" was technically possible, even if still risky.

4.3. European Trials in Galicia and Other Districts

Outside of Baku, the region of Galicia (then part of the Austro-Hungarian Empire, encompassing areas of modern Poland and Ukraine) had natural seepages. Residents had long collected the oil from surface pools. By the 1810s and 1820s, local entrepreneurs started using simple drilling rigs—akin to salt wells—to probe deeper deposits. While documentation is sparse, evidence suggests that a handful of early wells reached 30 or 40 meters, producing small but notable volumes of crude.

Some of this oil was sold in local markets as lamp fuel or used in rudimentary mechanical processes. The success varied widely, with frequent collapses of well shafts or infiltration by water. Nonetheless, these attempts confirmed that deeper geologic layers could be accessed if the driller had patience, decent tools, and a bit of luck.

5. The Role of Early Geologists and Surveyors

As discussed in earlier chapters, the late 18th and early 19th centuries witnessed the rise of geology as a distinct scientific discipline. Geologists took interest in identifying rock strata, understanding fossil records, and mapping mineral resources, including coal and metals. Oil seepages, though less central to mainstream geology at the time, still caught some attention.

5.1. Searching for "Indicators"

Geologists of this era, such as those working in the United States or parts of Europe, attempted to correlate surface seeps with specific rock formations. They guessed that oil might be more common in sedimentary basins, where ancient organic matter had been buried under pressure. Though their methods were rudimentary—lacking modern geophysical tools—they sought "indicators" like certain shale layers or bituminous rocks that might point to deeper oil.

This knowledge, while incomplete, offered practical drilling tips. Explorers or landowners who read geological pamphlets learned to look for outcrops of black shale or oily residue near rivers. If these signs appeared, it suggested that a deeper well might be worthwhile. Though not foolproof, these geological hints guided some of the more informed drilling ventures of the early 19th century.

5.2. Government Surveys and Resource Catalogs

In countries like the United States, new governmental institutions, such as state geological surveys, emerged in the 1820s and 1830s. Their purpose was to map resources—coal, iron ore, fertile farmland—and occasionally mention known oil seepages. While the main impetus for these surveys was often coal (seen as vital to fueling steam engines), the references to oil seepages gave local communities an official record that "there is petroleum in these hills."

Such official surveys lent credibility to the idea that oil was more than a random fluke. Landowners who discovered references to oil-bearing layers in these publications might become more confident about hiring a driller or investing in better equipment. Although the scale remained small, this was a key step in bridging academic geology with practical resource extraction.

6. Safety Hazards and Technical Limitations

Early 19th-century oil extraction was fraught with dangers and difficulties. Drilling technology, adapted from salt or water wells, was not designed for managing volatile hydrocarbons. Explosive pockets of gas, blowouts, or well collapses posed serious risks.

6.1. Gas Accumulations and Fires

Crude oil often came with natural gas dissolved or trapped in adjacent rock layers. If drillers unexpectedly hit a gas pocket, the escaping flow could be ignited by a spark, leading to dangerous fires. Wooden derricks and rope lines only made the hazard worse, as a single flame could reduce the rig to ashes. Early well operators sometimes tried to mitigate this by drilling slower or venting suspected gas pockets, but the science of gas detection was nearly nonexistent.

Additionally, storing crude oil in wooden barrels posed a fire hazard, especially if the site lacked proper ventilation or had open flames for illumination. Some small drilling outfits resorted to storing oil in pits lined with clay or rocks, but this led to contamination, evaporation of lighter fractions, and the ever-present risk of a spark igniting the open pool.

6.2. Well Collapse and Water Intrusion

Without steel or advanced well casings, early wells often collapsed once the hole went below stable soil layers. Drillers had to line the bore with wooden staves or hammered iron pipes, both prone to leakage or structural failure. Groundwater could seep in, contaminating the oil or making it impossible to gather a clean flow. Even a minor shift in the surrounding earth might break the rudimentary casing, forcing the entire operation to start over.

These setbacks meant that only the most persistent operators could push beyond shallow depths. Many gave up after repeated collapses or repeated dryness—failing to strike a significant reservoir. This high failure rate reinforced the notion that drilling for oil was speculative, best left to those with either strong financial backing or a personal willingness to endure repeated losses.

7. Community Responses to Early Drilling Efforts

When an individual or a small group decided to drill an oil well in the early 19th century, local communities sometimes reacted with curiosity, skepticism, or even hostility.

7.1. Social Views and Superstitions

Some residents worried about the potential for environmental disruption. They feared that drilling might release foul odors, pollute water supplies, or ignite unstoppable fires. Folktales and religious teachings sometimes warned against meddling with "black liquids" hidden underground. In deeply conservative areas, neighbors might openly discourage or even sabotage drilling attempts, believing them to be reckless.

On the other hand, a subset of locals saw drilling as a path to economic opportunity. If the well succeeded, it could bring revenue, attract tradespeople, and put the region "on the map." Local shops might sell supplies to the driller, blacksmiths might forge specialized bits, and so on. This tension between fear and hope characterized many early drilling endeavors.

7.2. Landowner Disputes and Claimed Royalties

As with other natural resources, land ownership determined who had the right to drill. Farmers who leased their fields to a driller might demand a share of the proceeds—an early form of royalty agreement. However, such contracts were informal and varied widely in fairness. If a neighbor believed the oil pool extended under his property, conflicts could arise.

In some regions, local courts started seeing cases about "subsurface rights," a concept still not clearly defined in law at the time. These early legal disputes over who owned the oil-laden strata beneath the land foreshadowed the far larger property battles that would emerge once oil became more lucrative in the later 19th century.

8. Technological Innovations: A Gradual Rise

Although the early 19th century was not yet an era of rapid industrial drilling, the wheels of innovation were slowly turning. Engineers and tinkerers in different regions introduced small improvements that, over time, would make deeper oil extraction more viable.

8.1. Better Drill Bits and Cables

Drill bits evolved from simple chisels to more specialized designs that could chip and crush rock efficiently. Blacksmiths forged bits with hardened steel edges, while some tried multi-pronged shapes. Similarly, rope and cable manufacturers experimented with stronger fibers, such as Manila hemp, or early wire rope in rare cases. While wire rope was expensive and not widely adopted until later, these incremental steps showed that the technology was inching forward.

8.2. Simple Pumping Mechanisms

Once a well reached an oil-bearing layer, extracting the fluid required a pump. Early systems used basic piston pumps powered by horses or waterwheels. Operators lowered a cylinder with a one-way valve and then pulled it up, capturing some oil each stroke. This was slow but better than scooping oil by hand. In some cases, drillers tried "sucker rods" attached to overhead beams, a precursor to the iconic walking beam mechanism used in later decades.

Though clumsy, these early pumps showed that once an oil flow was discovered, there were ways to bring it to the surface in moderate quantities—enough to fill a few barrels per day. For local consumption, that was often sufficient.

9. Rising Awareness of Petroleum's Potential

Despite the technical limitations, newspapers, trade publications, and local rumor mills gradually spread stories about successful or semi-successful drilling attempts. Travelers passing through regions like Pennsylvania or the Caucasus might write letters detailing a curious well that pumped out "barrels of black fluid." This slow but steady trickle of information encouraged a mindset shift.

9.1. Entrepreneurs' Growing Interest

Small-scale speculators and merchants began to wonder if oil could be turned into a more profitable venture. Some floated the idea of building refineries to make lamp oil that rivaled whale oil in quality—though the practical steps to refine oil were still rudimentary at the time (we will explore refining in the next chapter). Others discussed pooling money to fund deeper wells, hoping to corner a local market for cheaper fuel.

These entrepreneurial sparks remained localized. There was no widespread "oil mania" or rush, but pockets of enthusiasm emerged in resource-rich zones. Over the next few decades, these pockets would multiply, especially as more improvements in drilling and refining techniques appeared.

9.2. Calls for Scientific Collaboration

A handful of early 19th-century thinkers advocated for more systematic study of petroleum resources. They wrote pamphlets proposing that governments sponsor geological expeditions to identify prime drilling sites or that scientific societies collect data on well yields. While these proposals rarely translated into large-scale initiatives—governments were often preoccupied with other political or economic issues—they did plant the seeds for future structured collaboration.

As geology matured as a discipline, a few geologists started focusing specifically on coal, oil, and gas as related fossil fuels. They gave lectures describing how ancient organic matter, buried for millennia, might transform into combustible substances. This conceptual framework helped would-be drillers feel less like they were taking a wild guess and more like they were aligning with a plausible natural process.

10. International Exchanges and Observations

During the early 19th century, global travel and communication, while still slow, had improved since previous centuries. Sea routes connected continents, and diplomatic or commercial missions sometimes included savants or engineers who took notes on local industries.

10.1. Europeans Visiting Baku and the Middle East

Russian expansion around the Caspian Sea brought Western Europeans—engineers, soldiers, traders—into contact with the Baku oil fields. Some wrote detailed reports on the local pit-digging methods, describing the ingenuity of villagers who collected oil in large open pits and stored it in goatskin sacks or wooden barrels. Returning home, these visitors occasionally published articles or letters about what they saw, fueling the sense that oil extraction was more advanced in certain Eastern locales than many Westerners realized.

Travelers in the Ottoman-controlled regions also documented minor oil usage. They sometimes witnessed roadside vendors selling crude in clay pots, marketing it as a remedy or lamp fuel. While the West did not adopt these systems en masse, the cross-cultural exchange highlighted that petroleum was a globally recognized resource, albeit one still overshadowed by other commodities.

10.2. Americans Exploring Abroad

Likewise, a few American adventurers or merchants traveled overseas, taking note of drilling or refining practices that exceeded what they knew at home. They might share their findings in local newspapers upon returning to the States. For instance, an American who saw deeper wells near the Caspian might try to replicate those methods in Ohio or Pennsylvania.

This kind of transcontinental "idea swapping" was slow, often hampered by language barriers or inconsistent documentation. Still, it contributed a crucial element to the slow evolution of oil extraction technology. The more travelers compared notes, the closer the world edged toward standardized approaches.

11. Small but Significant Oil Industries Emerge

By the 1830s and 1840s, a few places around the globe began to exhibit early signs of an actual "oil industry"—though tiny compared to the booms of the later 19th century.

11.1. Baku's Expansion Under Russian Influence

As Russian authorities solidified control over the Caucasus, they sometimes granted licenses or privileges to entrepreneurs willing to improve oil extraction methods around Baku. Simple derricks, partial use of mechanical rigs, and government-endorsed trade connections led to a modest increase in output. While official records are incomplete, historians note that Baku was producing thousands of barrels annually by the mid-19th century—still small by later standards, but significant enough to supply nearby regions.

Some of this oil was used in lamps or for industrial tasks like leather processing, soap-making, or even experimental attempts at lighting city streets in certain Russian towns. The technology was still coarse, but it proved that a dedicated region could scale up from purely manual collection to something resembling an organized extraction operation.

11.2. Galicia's Local Wells and Lamp Oil Market

In Austrian-controlled Galicia, local entrepreneurs also pushed deeper wells, with some success in the 1830s and 1840s. They refined crude oil into lamp fuel (though primitive in quality), distributing it locally. Urban households, especially in towns that lacked easy access to whale oil or cheaper alternatives, found it a useful option despite its odor. Some local factories or workshops also used heavier oil fractions for lubrication.

Though overshadowed by other trade items in the Habsburg economy, Galicia's oil output gradually caught the attention of traveling merchants. By the eve of the broader 19th-century transitions, a small but stable market for "rock oil" lamp fuel had formed, foreshadowing a more extensive development in subsequent decades.

11.3. Sporadic American Ventures

In the United States, the East Coast and Southern states still relied mostly on whale oil and local resources, but in the western frontier regions—such as Pennsylvania, Ohio, and Kentucky—sporadic well drilling for salt occasionally struck oil. Entrepreneurs in these frontier zones recognized they could bottle and sell "Seneca Oil" or "American Medicinal Oil," sometimes touting it as a cure-all or a cheaper lamp fuel.

Although no large-scale drilling operation dominated the scene, a loose network of small wells, traveling salesmen, and local refineries (often just barns with distillation pots) began to emerge. This decentralized pattern set the stage for the more famous drilling initiatives that would erupt around the 1850s, culminating in the Drake Well (1859), typically acknowledged as the start of the "modern" oil industry in the United States.

12. Societal Perceptions and Future Outlook in the 1840s

By the 1840s, oil remained a niche resource, used in certain regions for specific purposes. However, broader social and economic shifts hinted at what was to come:

- **Population Growth**: Rapidly rising urban populations in Europe and North America demanded more lighting and lubrication for early factories.
- **Industrial Machinery**: As machinery in mines, mills, and early railways became more sophisticated, the need for high-quality lubricants increased.
- **Expensive Whale Oil**: Declining whale stocks and dangerous whaling voyages pushed up prices, causing some to look for cheaper alternatives.

In public discourse, scattered references emerged to the idea that petroleum might one day supplant whale oil if only it could be produced and refined more efficiently. Newspapers occasionally ran opinion pieces speculating on whether these black seepages might revolutionize lighting, though such articles were overshadowed by political news, wars, and more immediate concerns.

13. Challenges in Scaling Up

For all the interest and incremental progress, scaling up oil extraction in the early 19th century faced significant hurdles:

1. **Infrastructure**: Poor roads, limited canals, and expensive shipping costs made transporting bulky oil barrels difficult.
2. **Refining Know-How**: Basic refining techniques existed but were inconsistent, producing variable-quality lamp oil.
3. **Lack of Major Investors**: Without proven high returns, banks and wealthy financiers were hesitant to fund large drilling operations.
4. **Regulatory Uncertainty**: Laws regarding mineral rights, land leases, and safety standards were underdeveloped, adding legal and financial risks.

Consequently, most drilling remained small-scale, reliant on local capital and catering to local or regional demands. The next steps—refining breakthroughs and more sophisticated drilling tools—would be crucial in moving beyond these limitations.

14. Prelude to Innovation: The Stage Is Set

By the late 1840s and early 1850s, developments in various regions—Baku, Galicia, parts of North America—signaled that oil extraction was inching toward a turning point. The technology was not yet advanced enough for large-scale, well-controlled wells, but the ideas were circulating. Meanwhile, refining methods were slowly improving in workshops, as tinkerers discovered that partial distillation could yield cleaner-burning fuels.

In essence, the early 19th century laid a foundation of modest but critical progress:

- **Adaptation of salt drilling methods** for deeper oil extraction.
- **Local, small-scale usage** that tested the viability of commercial oil sales.
- **Growing scientific curiosity** about subterranean processes and how oil formed.
- **Incremental technical refinements**—stronger cables, better bits, primitive pumps—that hinted at deeper potential.

Though overshadowed by whaling, tallow, and other industries, oil extraction was no longer an oddity. It was a recognized possibility, carrying a kernel of promise for those with enough vision (and luck) to pursue it further.

CHAPTER 14: THE RISE OF EARLY OIL REFINING (19TH CENTURY)

Introduction

Even the best drilling techniques of the early 19th century could only produce small to moderate flows of crude oil. Moreover, raw petroleum—often thick, sulfurous, and loaded with impurities—burned poorly and smelled unpleasant. This harsh reality restricted its broader adoption for lighting or industrial uses. Yet, as the century wore on, a few pioneers discovered that by **refining** crude oil—separating it into lighter and heavier fractions—they could create far more appealing fuels and byproducts.

This chapter explores how early refining evolved in the 1800s, especially in areas such as Poland's Galicia, the United States, and Russia's Caucasus region. We will look at the rudimentary distillation apparatuses, the experiments of chemists and tinkerers, and the gradual development of a modest but influential market for "cleaner" oil products. These processes paved the way for a more systematic approach to petroleum, setting the stage for oil's eventual rise to prominence by the late 19th century.

1. Why Refining Mattered

Petroleum in its raw form posed several challenges for consumers:

1. **Strong Odors**: High sulfur content and other compounds made crude oil smell foul.
2. **Soot and Smoke**: When used in lamps, unrefined petroleum released thick smoke, blackening walls and damaging air quality.
3. **Instability**: Crude oil contained volatile fractions that could evaporate quickly or ignite unpredictably, posing safety hazards.

Refining offered solutions by splitting crude into distinct components, each with different boiling points and properties. Lighter fractions could serve as lamp fuel, heavier fractions found use as lubricants or roofing pitch, and the process helped reduce sulfur and other noxious elements. Although the earliest refineries were small-scale operations, they significantly enhanced oil's utility—and therefore its economic potential.

2. The Heritage of Distillation and Alchemy

Refining did not appear overnight. Its origins lay in centuries of **distillation** traditions, from medieval alchemy to the perfume-making and liquor industries. By the 18th century, chemists in Europe and elsewhere had become adept at separating liquids by boiling points, using **alembics** and **retorts**. Applying this knowledge to crude oil was a natural, if underexplored, progression.

2.1. Alchemical Roots

Medieval and Renaissance alchemists had long experimented with flammable oils, though mostly in pursuit of medicinal concoctions or the fabled philosopher's stone. While they seldom aimed to create fuel for lamps, they did record observations on how heating certain "earth oils" released lighter vapors. By the 19th century, some of these cryptic texts were reexamined, giving modern experimenters a hint of how to proceed.

2.2. Influence of Spirits and Perfumes

Distillers used pot stills to separate alcoholic spirits from fermented mash, while perfumers refined essential oils from flowers and herbs. Observing how these processes used controlled heat and condensation, early petroleum refiners borrowed the concept of fractionation. They recognized that if alcohol could be distilled off from a base liquid, so could the lighter hydrocarbons in crude oil—although the latter typically required higher temperatures and different materials to handle the thick, tar-like residues.

3. Early Refining Experiments in Europe

Across Europe, small workshops and individual chemists took the lead in figuring out how to refine petroleum. Their efforts were sporadic—driven by curiosity, local demand, or occasional government support.

3.1. Galicia's Pioneers

In Austrian-controlled Galicia, entrepreneurs extracting oil from shallow wells or seepages realized that raw crude's limited utility hampered profitability. By the 1820s and 1830s, a few small distillation setups appeared near oil-producing villages. Essentially large copper kettles or iron retorts, these devices heated the oil until the lighter fractions boiled off. Collectors at the top or side would capture these vapors and condense them in cooling coils.

The resulting "lamp oil," though still imperfect, burned more cleanly than raw crude. Local tradespeople discovered that if they repeated the distillation process or added certain chemicals (like lime or alkali) to remove impurities, they could reduce odor and smoke further. Over time, word spread that "Galician lamp oil" offered a cheaper alternative to whale oil for some households. Though volumes were small, these early refineries carved out a niche market regionally.

3.2. British and French Experimenters

Britain, with its advanced chemical industries, saw a few refiners dabbling in petroleum during the 1830s and 1840s. Most were overshadowed by the thriving coal gas and coal tar sector, which produced gas for street lighting and tar-based products for roofing. However, occasional articles in British journals mentioned "distilling rock oil" as a side project.

In France, some chemists inspired by Antoine Lavoisier's foundational work on elements and compounds recognized that petroleum contained various hydrocarbons. They tried to isolate these hydrocarbons in laboratory settings. A handful even patented "improved lamp oil" processes, though the commercial uptake was limited until bigger sources of crude became available.

3.3. Russia's Caucasus Region

Around Baku in the Caucasus, oil entrepreneurs realized that if they could refine the local crude, they might sell it more widely across the Russian Empire. By mid-century, small refineries emerged near the main extraction sites, producing lamp oil, lubricants, and heavier byproducts. This refined petroleum often traveled by caravan or small ships across the Caspian and then overland, supplying towns in southern Russia. These operations faced logistical hurdles—such as a lack of consistent containers and poor transport routes—but they did cultivate a stable local demand for refined oil.

4. North American Advancements in Refining

In the early United States, a patchwork of small-scale refining efforts sprang up alongside the slow rise of oil extraction in places like Pennsylvania, Ohio, and Kentucky.

4.1. Early "Tar Distilleries" and Patent Medicines

Before any major oil strike, some folks recognized that the same distillation apparatus used for making turpentine from pine resin or for producing whiskey could, in theory, handle crude oil. A few so-called "tar distilleries" dabbled in local crude, separating out a thin fraction marketed as lamp oil or a heavier fraction used as a protective coating for fences and roofs.

Meanwhile, traveling salesmen discovered that lightly refined oil, sometimes diluted with herbs, could be sold as a patent medicine—"Seneca Oil," "American Medicinal Oil," or similar brand names. While these concoctions had dubious medical value, they turned a profit and kept refining knowledge alive, ensuring a minimal but ongoing interest in how to process petroleum.

4.2. Workshop-Based Innovations

In and around western Pennsylvania—well before the famous Drake Well of 1859—tinkerers set up backyard or barn-based stills. They recognized that a single pass of distillation yielded a rough lamp fuel, while multiple passes improved clarity and smell. Some added bleaching clay or heated the oil with an alkali, attempting to neutralize acidic compounds.

These refiners sold their lamp oil in local general stores or directly to neighbors. Stories exist of certain homesteads using nothing but refined "rock oil" for lamps by the 1840s, simply because it was cheaper than shipped-in whale oil. These workshop operations were small, but they formed a knowledge network—neighbors swapped tips, borrowed each other's still designs, and refined the process step by step.

5. Distillation Apparatus and Techniques

Most early 19th-century refining setups shared similar features, adapted from existing distillation technology:

1. **Boiler/Retort**: A strong metal container where crude oil was heated over a fire.
2. **Condenser Coil**: A long metal or copper coil cooled by water or air, where hot vapors condensed back into liquid.
3. **Collection Vessel**: Often a barrel or metal container that collected the distillate.

5.1. Batch Distillation

In the earliest days, refiners performed **batch distillation**—filling the still with a certain amount of crude, heating it until the lighter fractions boiled off, then collecting them. Operators watched the temperature carefully (using rudimentary thermometers) to gauge when certain fractions emerged. The process was repeated until only heavy tar remained at the bottom, which sometimes was used for roofing or pitch.

Refiners learned by trial and error what temperature ranges produced the best lamp fuel. They noticed that heating crude too hot or too fast created sooty residues and risked fires. A slow, steady heat tended to yield a more uniform product. Still, controlling temperature precisely was challenging with basic wood or coal fires.

5.2. Fractionation and Multiple Passes

Refiners discovered that repeating the distillation multiple times improved quality. A second or third pass removed more impurities and lighter volatile elements that caused odor and smoke. Adding a chemical wash—like an alkaline solution—between passes sometimes further reduced sulfur.

Though each repeated pass improved the oil, it also increased labor and fuel costs. Many small operators sought a balance: create a decently clean fuel at minimal expense. This middle ground meant that early lamp oils were not pristine, but they were good enough to compete with, or at least supplement, existing lamp fuels. Over time, refiners who excelled at multi-pass distillation earned a local reputation for producing high-grade "burning fluid," attracting repeat customers.

6. The Emergence of Specialized Refinery Owners

While some refiners were farmers or blacksmiths doing side work, a few dedicated entrepreneurs emerged in the 1840s and 1850s, establishing proto-refineries. They built larger boilers, installed multiple stills, and attempted to produce a consistent, branded lamp oil for sale in nearby towns. This was particularly true in the following locales:

- **Galicia**: Larger refineries near oil wells sold "Galician lamp oil" beyond local villages, reaching urban markets in parts of the Austro-Hungarian Empire.
- **Baku Region**: Some local owners formed partnerships with merchants to refine oil in small factories, shipping the product across the Caucasus.
- **North America**: In scattered pockets of Pennsylvania and Ohio, a few entrepreneurs built more structured refining operations, sometimes employing several workers.

6.1. Marketing and Distribution

These proto-refiners often faced the challenge of convincing the public that their lamp oil was safe and efficient. Whale oil had a strong reputation, and older forms of lighting were well-established. Refiners might demonstrate their product in a local inn or tavern, burning it in a lamp to show fewer fumes or a brighter flame. If successful, they could secure standing orders from local shops or factories needing lubrication.

To distinguish their output, some gave it brand-like names or attached seals guaranteeing purity. This early marketing hinted at the modern concept of brand recognition, though on a tiny scale.

6.2. Economic Ups and Downs

The oil refining business in the early 19th century was precarious. One year, whale oil prices might spike, creating a surge of interest in petroleum-based alternatives. The next year, a good whaling season or an economic slump might reduce demand, driving refineries into financial trouble. Many refined products were sold locally, limiting growth. Only those with enough capital to endure market fluctuations survived.

Additionally, shipping refined oil over poor roads or via slow river transport raised costs. It was simpler to supply towns within a short radius. This geographic constraint meant small refineries often popped up near well sites, forging tight local links between drilling and refining operations.

7. Science and Chemistry Evolve: The Mid-Century Breakthroughs

As the 19th century progressed, the field of **organic chemistry** took shape. Researchers studying carbon-based compounds realized that petroleum held a complex mixture of hydrocarbons. Their findings opened doors to more methodical refining approaches.

7.1. Identifying Hydrocarbon Families

Scientists like Jean-Baptiste Dumas in France and others across Europe explored how carbon and hydrogen bonded in various ratios. While they did not fully grasp the nuanced structures we know today, they recognized patterns: lighter fractions were more volatile and suitable for lighting, while heavier fractions could serve as lubricants or wax-like products.

Chemistry journals occasionally featured papers analyzing oil samples from different regions, noting differences in density, color, and boiling range. This data guided refiners who used trial and error less blindly, starting to keep logs of which "cuts" burned best in lamps. Although still rudimentary, this synergy between theoretical chemistry and practical refining marked a key turning point—oil refining was becoming less guesswork and more science.

7.2. The Role of Educational Institutions

A handful of universities or technical schools, especially in Germany and France, offered specialized classes on industrial chemistry. Students there sometimes studied real petroleum samples. These academically trained individuals might go on to manage or consult for early refineries, sharing structured knowledge about fractionation, temperature control, and purification methods.

Though far from universal, this slow infiltration of formal chemical training into refineries helped standardize processes, reduce accidents, and yield more consistent products. By the 1850s, some of the better-run refineries in places like Eastern Europe or the eastern United States were increasingly directed by individuals with at least some formal scientific background.

8. Products of Early Oil Refining

Refiners in the early to mid-1800s typically divided crude oil into several categories of output:

1. **Lamp Oil or Burning Fluid**: The most commercially important fraction. It needed to be relatively clear and low in sulfur.
2. **Lubricating Oils**: Slightly heavier fractions used on machinery, wagon axles, or factory equipment, though these were less in demand before widespread industrial mechanization.
3. **Residues/Tar**: Thick leftover material at the bottom of the still, used for roofing, caulking ships, or road patching.

8.1. Lamp Oil Quality and Competition

Lamp oil was the critical product if refiners hoped to challenge whale oil. Consumers wanted an oil that ignited easily but did not produce heavy smoke. Achieving this balance was tricky. Refiners discovered that certain crude sources produced better lamp fractions than others: for instance, oil from one region might have fewer sulfur compounds, making it more suitable for lighting.

Some operators tried bleaching their lamp oil with sunlight or chemical additives, improving clarity and reducing odor. Others used multiple distillation passes, labeling the final product as "double-distilled" or "triple-distilled." Price remained a factor: these extra steps raised costs, so refiners had to gauge whether the local market would pay more for a cleaner burn.

8.2. Rise of Greases and Lubricants

As industrial machinery slowly spread, heavier petroleum fractions found new uses as lubricants. While early steam engines and factory machines often relied on animal fats or vegetable oils, experimenters tested thicker petroleum-based oils for bearings or gears. Some found that these oils resisted breakdown at higher temperatures, though they could be sticky or contain abrasive impurities if not well-refined.

Thus, a minor but growing subset of refiners specialized in producing "engine oils," marketing them to mills or rail workshops. By mid-century, the concept that petroleum could lubricate complex metal parts was gaining acceptance, foreshadowing a massive future demand once industrialization intensified.

9. Barriers to Wider Adoption of Refined Oil

Despite the clear advantages of refining, multiple factors still limited refined petroleum's widespread acceptance:

1. **Cost**: Proper distillation required equipment, fuel, and skilled labor, making the final product expensive compared to established fuels.
2. **Distribution**: Many communities lacked easy access to refineries. Shipping refined oil over distances was cumbersome without robust transport links.
3. **Quality Inconsistencies**: Small refineries varied in skill, leading to a range of product qualities. A bad batch could sour public opinion.
4. **Entrenched Competitors**: Whale oil, tallow, coal gas, and even pine-derived turpentine had well-established supply chains and customer loyalty.

Nevertheless, each incremental improvement in refining technology or distribution chipped away at these obstacles, setting the stage for expansions once certain breakthroughs—like more abundant crude supply—would occur.

10. Case Studies: Refining Centers Before 1859

10.1. Galicia's "Workshops"

By the 1840s, a handful of Galician operators established modest refineries near small towns with notable oil wells. One or two had upward of a dozen employees, operating multiple stills in a row. They typically sold lamp oil in wooden barrels or metal containers to merchants who distributed it in local markets. While overshadowed by large agricultural exports like grains, Galician refined oil carved a niche among rural households that struggled to afford whale oil.

Reports from travelers in the region sometimes mention these "oily" villages, describing the strong smell of distillation day and night. Skilled workers stoked coal or wood fires under the stills, carefully monitoring temperatures. The daily output might be measured in tens of gallons—enough to supply a radius of a few dozen miles. Profits were modest, but stable in regions with limited access to alternative fuels.

10.2. Baku's Early Refineries

Around Baku, where oil seeps were abundant, local refiners initially used open-air boiling methods to isolate lighter fractions. Over time, some adopted enclosed stills, reducing evaporation loss and the risk of ignition. The refined product found markets in adjacent cities and for powering small industries like soap-making or glass production.

As Russian influence in the Caucasus grew, certain officials encouraged entrepreneurs to upgrade these refineries. Government documents from the 1840s occasionally mention state interest in "improved methods of oil purification," though large-scale modernization would only come later. Still, the seeds were planted, and Baku's refiners demonstrated that consistent refining practices could expand an oil product's reach far beyond local usage.

10.3. North America's Patchwork Scene

In the United States, prior to the Drake Well of 1859, refineries were often ephemeral: a barn or shed with a few stills, reliant on neighboring seeps or small wells. Owners operated them intermittently, shutting down if crude supplies ran low or if local demand dipped. Some products were sold as lamp oil, others as patent medicines.

Areas like western Pennsylvania gained a reputation for "Rock Oil" that could be refined into half-decent lamp fuel. Local storekeepers might stock jars or small barrels. People living in out-of-the-way settlements sometimes found it cheaper than shipping in whale oil from coastal cities. Yet the scale remained extremely modest—no cohesive "industry" existed, just scattered pockets of enterprise.

11. The Social and Cultural Dimensions of Early Refining

Refining was more than a mere technical process; it also had social and cultural impacts.

11.1. Community Nuisance or Economic Opportunity?

The smell and smoke associated with small refineries could be intense, causing neighbors to complain if the site was near residences or farmland. Some local authorities tried to zone refining operations away from busy areas, citing health or fire concerns. Conversely, communities that embraced refining might see it as a source of local pride—evidence of modernization and self-sufficiency.

Refinery owners who offered stable wages or bought local crude from farmers' wells boosted the area's economy. This economic ripple could lead local leaders to protect the operation, even if it was occasionally disruptive. Thus, local sentiment about refining varied widely, shaped by the perceived balance between inconvenience and benefit.

11.2. Early Labor Practices

Workers in early refineries needed to be comfortable with the stench and hazards. Many were unskilled laborers or farmers supplementing income, though some refineries hired a "chief distiller" with more specialized knowledge. Conditions were tough: open fires, hot retorts, risk of burns or inhalation of fumes. Wages were typically modest, but a steady job in a rural setting offered some security.

Labor disputes or formal organization were rare at this stage. Most refineries were too small for large-scale labor movements. Instead, relationships were personal, with owners sometimes living near or among their employees. This informal dynamic kept many such businesses stable but also limited their expansion potential—scaling up would require more structured operations and capital.

12. Steps Toward a More Systematic Industry

By the early to mid-1850s, refining techniques in scattered places had grown more reliable, even if not advanced by modern standards. A convergence of factors now set the stage for an industry poised to take off if more abundant crude supplies became available:

1. **Rudimentary Standardization**: Repeated distillation or chemical washes were recognized methods, albeit with local variations.
2. **Market Awareness**: Households knew refined petroleum existed, sometimes preferring it over poorly refined alternatives.
3. **Slow Tech Improvements**: Better still designs and safer, enclosed burners reduced the risks of fire and explosions.
4. **Chemistry's Evolution**: New findings in organic chemistry gradually illuminated the structures of hydrocarbons, hinting at how refining could become more precise.

At this juncture, refining was akin to a set of craft practices, shaped by each refiner's skill and resources. There was no single blueprint or large corporate model. Yet the pieces were in place for rapid scaling once an event or innovation broke the supply and distribution bottlenecks. That event would come, famously, in the form of more productive drilling that revolutionized the quantity of oil on the market.

13. Interplay with Other Emerging Fuels

While refining improved, other energy technologies also advanced in the early to mid-19th century—most notably coal gas and coal-derived lighting solutions. Gas lighting spread in European and American cities, providing bright streetlamps and commercial lighting. This stiff competition forced petroleum refiners to focus on rural or small-town consumers, or niche industrial uses.

Additionally, experiments with compressed gas or camphene (a mix of turpentine and alcohol) found some success in lamps. These alternatives meant that refined petroleum never enjoyed a total monopoly, even locally. The multi-fuel environment kept refiners on their toes, encouraging them to keep improving product quality and cost efficiency.

14. Transition to Larger Opportunities: Late 1850s Outlook

By the late 1850s, those refining oil realized that bigger changes loomed on the horizon:

- **Rumors of "better wells"** in North America started circulating, as drilling for oil (not salt) became more deliberate.
- **Improved transport** options like canals and railroads began linking inland refineries to broader markets.
- **Growing urban populations** demanded better, cheaper lighting, especially at night for shops, streets, and factories.

Some savvy entrepreneurs recognized that if they could secure a stable, higher-volume crude source, they could expand refining dramatically. This anticipation set the stage for the big leaps that occurred post-1859. While that year's Drake Well is typically cited as the spark of the modern oil era in the U.S., the refining techniques built up over previous decades allowed newly discovered oil to be quickly turned into marketable fuels.

CHAPTER 15: OIL AND TRANSPORTATION SHIFTS OF THE 19TH CENTURY

Introduction

The 19th century was a period of dramatic changes in how people and goods moved across land and water. The expansion of railroads, the refinement of steamship technology, and the ongoing use of horse-drawn vehicles transformed economies and societies worldwide. As these shifts unfolded, oil began to play a small but significant role—primarily in lubrication, maintenance, and occasionally as a supplemental fuel. Although coal remained the dominant energy source for large steam engines, and whale oil or tallow persisted for certain applications, many inventors and entrepreneurs recognized that petroleum had unique properties that could prove valuable for transportation.

In this chapter, we will explore how early oil products intersected with the evolving transportation methods of the 19th century. We will see how trains, ships, and other vehicles gradually integrated petroleum-based lubricants, how some visionaries dreamed of using oil as a fuel, and how improvements in drilling and refining made these ideas increasingly feasible. While oil did not yet dominate the transportation sector, the seeds of future transformations were planted during this century. These developments laid a foundation for larger changes that would follow in the closing decades of the 19th century and beyond—though our focus here remains strictly on the 1800s, stopping before modern times.

1. The Global Transportation Landscape in the Early 1800s

1.1. Dominance of Horse-Drawn and Wind-Powered Travel

At the dawn of the 19th century, travel across land depended mostly on horse-drawn wagons, coaches, or, in many regions, simple ox-carts. On rivers and seas, wind-powered sailing vessels still carried the bulk of global trade. Steam engines, though pioneered in the late 18th century, were still in their infancy. Early steamboats began appearing on major rivers in Europe and North America, but they faced mechanical challenges, high coal consumption, and maintenance demands that slowed their spread.

Within this environment, oils and lubricants were needed to keep wheels turning smoothly, reduce friction in simple machinery, and maintain rudimentary gear systems. Animal-based greases, such as tallow, and plant oils, like linseed or castor oil, served these purposes. Whale oil was also common for lubrication in maritime contexts. While traces of petroleum might have found their way into local uses—such as greasing wagon axles in Pennsylvania or the Caucasus region—few people considered it a widespread transport solution at this early stage.

1.2. Slow Emergence of Steam Transport

As the 19th century progressed, steam-powered vehicles grew more common. Pioneering locomotives in Britain demonstrated that iron rails and steam engines could haul heavy loads at unprecedented speeds. Similarly, improved steamboats plied major rivers such as the Mississippi in North America, or the Rhine in Europe. Coal was the essential fuel for these engines, and whale or other natural oils remained the favored lubricants for moving parts. Petroleum was little known to these early transport pioneers, except in small pockets where local experimenters found ways to exploit seepages.

Despite this general reliance on older fuels and lubricants, the seeds of curiosity about petroleum were already present. Some inventors recognized that heavier fractions from crude oil were more heat-resistant than animal fats, hinting they might excel as lubricants in steam-driven machinery. Others noted that certain lighter petroleum distillates seemed to burn with intense heat, raising the question of whether oil could someday power steam boilers. These ideas would take decades to mature, but the impetus for searching beyond traditional materials was already visible.

2. The Growth of Railways and the Need for Better Lubrication

2.1. Early Locomotives and Friction Challenges

The first commercially viable railways, beginning in Britain around the 1820s and spreading globally throughout the century, encountered serious friction and wear issues. Steam locomotives required numerous moving parts—pistons, connecting rods, wheel bearings, and more. The intense heat and pressure inside these mechanisms demanded lubricants capable of withstanding higher temperatures without breaking down into gummy residues.

Traditional lubricants, such as tallow or lard, often melted or carbonized under the locomotive's operating conditions. While these greases worked decently for slow-moving carriage axles, high-speed rail travel called for more specialized solutions. Engineers scrambled to find or develop lubricants that would not degrade so quickly, cause excessive friction, or corrode metal surfaces.

2.2. Experiments with Animal and Vegetable Oils

In the first half of the 19th century, railway operators experimented with blends of tallow, olive oil, rapeseed oil, or castor oil. Each had unique drawbacks. Tallow grew rancid, vegetable oils sometimes oxidized into sticky varnishes, and whale oil—though it performed better under moderate heat—was expensive and threatened by uncertain supply. These limitations spurred investigations into new options.

Some railway workshops, hearing rumors of "rock oil" that withstood heat more effectively, tested small batches on bearings or piston rods. Feedback varied: early, poorly refined petroleum often contained impurities that left abrasive deposits or emitted foul odors. Yet a handful of engineers realized that if they could obtain a purer form of heavy petroleum fraction, it might outperform tallow under hot, fast-moving conditions.

2.3. Gradual Introduction of Petroleum-Based Lubricants

As the century wore on, improvements in oil refining (discussed in the previous chapter) produced more consistent lubricating oils. By the 1840s and 1850s, a small but growing subset of rail companies began buying these heavier petroleum fractions for specific locomotive applications—particularly rod bearings or gearboxes. Because the supply of refined petroleum was still limited, and distribution networks were rudimentary, only well-funded lines or those near emerging oil regions could adopt such lubricants regularly.

Despite these constraints, the success stories of certain railway workshops demonstrated that refined petroleum oils could extend the intervals between re-lubrication and reduce wear on metal parts. This built quiet momentum toward broader acceptance of petroleum-based lubricants. Word of mouth among locomotive engineers, combined with scattered technical articles in railway journals, kept interest alive. Even if petroleum never replaced tallow entirely in mid-19th-century railways, it staked out an important foothold.

3. Steamships and Marine Lubrication

3.1. The Rise of Steam Navigation

Parallel to the rail revolution, steamships advanced swiftly in the 19th century. Early paddlewheel steamers plied major rivers and coastal routes, gradually transitioning to iron-hulled vessels capable of crossing oceans under steam power. Coal bunkers fueled massive boilers, but these engines also demanded reliable lubrication for crankshafts, rod journals, and piston cylinders. Marine conditions introduced additional challenges such as corrosion from saltwater and extreme temperature variations.

Marine engineers initially relied on whale oil or tallow for lubrication, just as they had for rigging and mechanical parts on sailing ships. But they too encountered the limits of these traditional products at high temperatures. The thick sludge formed by overheated tallow or whale oil could clog machinery, leading to breakdowns mid-voyage. This problem, paired with increasing whale oil costs as whaling fleets ventured into distant oceans, prompted captains and shipowners to look for alternatives.

3.2. Early Trials with Petroleum at Sea

By the 1830s and 1840s, a handful of ship engineers tested small quantities of petroleum-based lubricants. They discovered that certain heavier distillates, if well-refined, remained stable at the high temperatures inside a steamer's engine. These oils did not degrade into gummy residues as quickly, though they sometimes carried a strong odor or contained corrosive elements if refinement was incomplete.

The biggest challenge was supply. Ships needed consistent deliveries of lubricants at major ports. Petroleum refineries were few, and distribution by sea was costly. Some vessels operating near the Caspian Sea or the Black Sea—where Baku's oil was more accessible—adopted petroleum-based lubricants earlier than those on the Atlantic. Likewise, ships that frequented the American East Coast might experiment with refined oils from small Pennsylvania or New York operations if they found a local agent.

These scattered successes fed into maritime trade gossip, with captains swapping notes on which oils performed best. The general consensus was that if a ship could reliably source a stable, heavy petroleum fraction, it could reduce lubrication failures and extend maintenance intervals. However, given the limited volume of refined oil available, widespread adoption remained modest until later in the century.

3.3. Influence on Shipbuilding

Marine engineers tinkered with design choices influenced by the improved lubricants. For instance, piston rods and bearings could be built with tighter tolerances if the oil quality was consistent. Some shipyards began specifying "petroleum-based lubricants only" in their guidelines for new steamers, especially if owners believed in the product's superior performance.

Still, the transition was piecemeal. Many shipping lines stuck to proven whale oil or blended solutions to avoid risking engine malfunctions at sea. Yet each success story—like a steamship crossing an ocean with minimal bearing wear—demonstrated that petroleum could match or surpass traditional lubricants in certain conditions. This shift, though gradual, laid the groundwork for more significant changes once refining technology advanced further and drilling yields increased.

4. Road Transport: Wagons, Coaches, and the Slow Adoption of Oil

4.1. Persistence of Animal Greases for Wagons

Horse-drawn wagons, stagecoaches, and carriages crisscrossed both rural and urban roads throughout the 19th century. Their axles, wheels, and pivot points required frequent lubrication to reduce friction and prevent metal-on-metal wear. Families and small-time haulers typically used animal-based greases—tallow, lard, or a blend of animal and plant oils—because they were cheap, readily available, and understood.

In certain areas near oil seepages—such as parts of Pennsylvania or the Carpathian region—farmers replaced or supplemented these greases with thick crude or partially refined oils. Although pungent, these petroleum products could provide smoother performance under heat and stress, especially during long-distance hauls or in mountainous terrain where brakes and wheels were subjected to extra friction. Nonetheless, this practice remained highly localized. The average wagoner was unlikely to seek out specialized petroleum lubricants unless local tradition and supply made it practical.

4.2. Early Stagecoach Companies and Trials

Stagecoach companies that operated extensive routes, especially in Europe and North America, began paying more attention to maintenance costs and reliability. Broken axles or overheated wheels could mean costly delays and unhappy passengers. Hearing rumors of "rock oil" that performed well under stress, certain stagecoach operators tested small volumes of refined lubricants.

Some discovered that a heavier petroleum fraction mixed with small amounts of tallow produced a stable grease that resisted thinning in hot weather. This prevented wheels from seizing during long descents or bumpy roads. Though these improvements were incremental, they contributed to a slow but steady acceptance of petroleum-based solutions. Still, adoption was uneven, depending on proximity to refining centers and the willingness of stagecoach owners to pay a premium for more advanced lubricants.

4.3. Local Variations in Acceptance

Cultural and economic factors shaped whether local road transport embraced petroleum lubricants. In wealthier regions with busier commercial routes, coach lines or freight companies were more likely to afford experimental products. In poorer or isolated areas, people stuck to time-honored greases. Government mail services, which often contracted stagecoach operations to deliver post, sometimes mandated higher standards of vehicle upkeep, indirectly spurring interest in advanced lubricants.

These patchwork patterns meant that by mid-century, a traveler could find one district still using purely animal-based axle greases, while another boasted of new petroleum-laced solutions. Cumulatively, these small-scale adoptions reinforced the notion that oil products were not just for local oddities but had potential to serve broader transportation needs.

5. Attempts to Use Petroleum as a Primary Fuel

5.1. The Dominance of Coal in Steam Boilers

Throughout the 19th century, coal reigned as the principal fuel for stationary steam engines, locomotives, and marine boilers. It was abundant, relatively cheap in industrialized nations, and the technology for burning coal in large furnaces was well understood. However, a few bold experimenters wondered if refined oil could power a steam boiler more cleanly or efficiently—especially if it produced less ash or could reach higher temperatures faster.

Early trials showed that while oil could indeed burn with high heat, feeding a continuous liquid fuel into a boiler was more complex than shoveling coal. Devices such as oil injectors or atomizers were largely unknown or in embryonic development. Uncontrolled ignition also posed significant risks. Moreover, the limited supply of refined oil and its higher cost compared to coal made the proposition uneconomical for mainstream railway or maritime operators.

5.2. Occasional Firebox Experiments

In certain locales where crude or partially refined oil was cheaply available—like near Baku, or small pockets of North America—individual inventors rigged up rudimentary devices to burn oil in a firebox. They might drip oil onto hot coals or attempt to spray it through a nozzle. Some achieved enough success to heat small stationary boilers used for sawmills or salt evaporation. But these contraptions were usually homemade, prone to malfunction, and demanded constant operator attention to avoid flare-ups.

As news of these experiments spread, a tiny circle of enthusiasts proposed that oil-based steam engines might eventually challenge coal-fired systems. However, mainstream engineers dismissed the idea as impractical without a large, stable supply of refined oil. Indeed, prior to the late 19th century, no major railway or steamship fully converted to oil firing. Coal remained king, with petroleum's role in fueling heavy transport negligible. That said, these early attempts foreshadowed the future potential of liquid fuels once technology and supply chains caught up.

5.3. Rumblings of Internal Combustion Ideas

While well beyond this book's primary historical scope, it is worth noting that a handful of 19th-century inventors toyed with concepts that resembled internal combustion engines, albeit in very primitive forms. They recognized that vaporized oil or gas could theoretically drive pistons directly, bypassing steam. French, German, and British patent offices recorded scattered submissions for "explosive engines" or "oil vapor engines," though none achieved significant commercial traction in the mid-1800s.

These prototypes typically faced issues like poor vaporization, uncontrolled explosions, and a lack of suitable ignition systems. Furthermore, stable refined petroleum fractions were scarce, limiting the prototypes' operational reliability. While these early explorations did not reshape transport in the 19th century, they sowed seeds for bigger leaps in the decades ahead. In the context of this chapter, it suffices to say that petroleum-based fuels had begun to enter the imaginative realm of engine developers, even if no practical solutions emerged yet.

6. Roads, Infrastructure, and Bitumen for Surfaces

6.1. Highway Improvements in the 19th Century

Alongside the dramatic shifts in rail and maritime transport, road infrastructure also underwent transformations. Governments in many industrializing nations recognized the need for better roads to facilitate commerce and mail delivery. Turnpike trusts in Britain, macadamized roads in the United States, and new highways on the European continent spurred an increase in paving and surface maintenance.

While stone, gravel, and macadam were common, a minority of engineers experimented with bituminous binders—often derived from crude petroleum residues or natural asphalt. These substances could hold gravel together, creating more durable, water-resistant surfaces. The famous Pitch Lake in Trinidad and tar from the Middle East had historically been used in maritime caulking; now, some civil engineers saw potential in applying these sticky materials to roads.

6.2. Early "Tar Road" Experiments

By mid-century, local authorities in certain towns or cities initiated small-scale trials of mixing petroleum residue with sand or gravel to form a rudimentary asphalt. If the region had easy access to refining byproducts, the cost might be competitive with alternatives. Observers noted that such surfaces resisted rain damage better and reduced dust on busy streets. However, some forms of crude tar or pitch gave off a strong odor, especially in hot weather, causing public complaints.

In the United States, a few big cities tested patches of "tar-bound macadam," with limited success. Europe's interest in asphalt roads was more prominent by the late 1800s, but these earlier efforts laid the groundwork. They also created a modest demand for the heavier residues left over from oil refining, turning what might have been waste into a profitable commodity. Although not directly tied to the vehicles themselves, these bituminous roads were part of the broader synergy between oil products and the era's evolving transportation network.

6.3. Impact on Horse-Drawn Traffic

Improved road surfaces indirectly benefited horse-drawn transport by reducing ruts and potholes, hence cutting repair costs for wagon and carriage owners. Additionally, a smoother ride meant less friction, further diminishing the need for frequent axle greasing. This synergy exemplifies how a single oil byproduct—bitumen—could influence daily travel patterns, even if the direct link to "petroleum as a transport fuel" was not yet established.

Although these bitumen roads remained a rarity in most regions, the concept demonstrated that oil's relevance to 19th-century transportation could extend beyond engines and lubricants. Road-building challenges aligned with petroleum-based solutions, hinting at the wide-ranging versatility that oil would later exhibit on a grander scale.

7. Cultural and Economic Drivers of Petroleum Use in Transport

7.1. Rising Consumer Demand for Efficiency

As industrialization broadened consumer markets, people expected faster, more reliable transportation—both for themselves and for goods. Travel times that once took days or weeks were shortened by trains and steamboats, fueling an appetite for additional improvements. Many transport companies realized that dependable machinery with minimal breakdowns was key to staying competitive.

Consequently, they welcomed lubricants that reduced maintenance intervals and unexpected failures. In this sense, a high-quality petroleum product, although more expensive than basic lard, could pay off by minimizing downtime and preventing catastrophic failures mid-journey. This practical calculus undergirded the slow but meaningful shift to petroleum-based lubricants in advanced transport networks.

7.2. Variations by Region and Industrial Growth

Wherever industrial activity was highest—like Britain's busy railway lines, the manufacturing corridors of the United States, or continental Europe's major trade hubs—the impetus to experiment with new lubricants was stronger. In more rural or less industrialized areas, the pace was slower. For instance, a small carriage service in a remote province might continue using pig-fat grease for decades, seeing no reason to adopt an "exotic" oil product.

Economic factors such as disposable income, availability of local refining, and the presence of robust transport links shaped how quickly petroleum infiltrated local supply chains. Regions near oil wells and small refineries, like Pennsylvania or Baku, had an advantage: they could supply cheaper petroleum to local railroads or steamship companies, fostering early adoption. In distant locations lacking direct access, the cost of importing refined oil might offset any performance benefits.

8. The Role of Inventors, Entrepreneurs, and Tinkerers

8.1. Workshops and Small Engineering Firms

Rather than top-down mandates from large corporations, the impetus for introducing petroleum products often came from grassroots-level workshops or small engineering firms. Tinkerers who specialized in locomotive components or marine engines might buy a barrel of experimental oil, test it on a spare engine, and note the results. If the trial succeeded, word spread quickly among professional circles.

Local entrepreneurs, seeing potential profit, sometimes partnered with early oil refiners to create specialized products labeled "locomotive bearing oil" or "marine engine lubricant." They printed basic advertising in trade journals or circulated flyers at engineering conventions. These micro-level efforts collectively spurred the slow acceptance of petroleum in transport maintenance.

8.2. Cross-Pollination of Ideas

Nineteenth-century trade fairs, expositions, and engineering societies played a moderate role in sharing technical insights. Engineers from different countries might see a demonstration of a new locomotive bearing grease or read about how a steamer completed a route with fewer breakdowns thanks to a certain petroleum blend. Although communication was slower than in modern times, such cross-pollination ensured that no single region entirely monopolized the concept of using oil in transport.

Occasionally, local governments offered modest prizes or recognition for inventors who reduced maintenance costs or improved safety. Some of these awards went to individuals who championed petroleum lubricants. This recognition, though not lavish, gave them credibility and encouraged further refinement of their methods. Over time, these incremental innovations would accumulate, shaping a more formal market for petroleum-based transport solutions.

9. Environmental and Safety Considerations

9.1. Early Awareness of Spills and Fumes

Most 19th-century transport operators did not articulate "environmental concerns" in the modern sense, but they did worry about the mess and hazards of petroleum. Spilling refined oil on a ship's deck or in a locomotive yard was slippery and potentially flammable. If a barrel leaked along a railway platform, it could create a noxious odor, discouraging passengers.

Some railway companies or shipping lines set basic rules for storing and handling oil barrels. They might require them to be kept away from open flames, use robust containers, or station them in well-ventilated areas. These rudimentary precautions, while not driven by a conscious environmental ethos, at least recognized the practical hazards and tried to mitigate immediate risks.

9.2. Fire Hazards in Engine Rooms

Steamboat engine rooms and locomotive sheds, full of hot machinery and coal-fired boilers, were tinderboxes for potential fires. Introducing petroleum-based products added a new element of flammability if leaks occurred. Hence, some operators hesitated to adopt oil lubricants, fearing an accident. Over time, improved container design and more careful application practices reduced these risks, but the fear lingered, slowing adoption in more conservative operations.

When small or moderate fires did break out due to spilled oil, local newspapers sometimes sensationalized these events, framing petroleum as dangerously volatile. This negative publicity reinforced caution among the public and industry. Only consistent demonstration of safe usage, combined with the tangible benefits in lubrication, gradually overcame such concerns.

10. Late 19th-Century Shifts on the Horizon

10.1. The Impending Oil Boom

By the 1850s, rumors swirled of larger oil finds, especially in North America, where some deeper wells had begun producing a steady flow of crude. Entrepreneurs speculated that if large volumes of affordable petroleum reached the market, it could reshape everything from lighting to transport lubrication. Refineries, still mostly small and localized, might scale up if they had a reliable feedstock.

For transport sectors, the promise of cheaper, higher-quality lubricants was appealing. Engineers and managers started paying attention to any news about new wells or refining breakthroughs. Although no one predicted that oil would one day eclipse whale oil or become a global strategic commodity, the sense of potential was growing. Some forward-thinking railway executives quietly laid plans to adopt petroleum lubricants more broadly if the supply stabilized.

10.2. Technological Synergy

Steady refinements in locomotive design, marine engines, and road vehicles interacted in subtle ways with improving oil products. For instance, if locomotive cylinders were built to handle higher pressures, they needed a more advanced lubricant. This further spurred refiners to refine heavier fractions or add chemical treatments. Meanwhile, the search for better road surfaces spurred research into asphalt mixtures, linking transport infrastructure with petroleum byproducts.

All these trends hinted that the latter half of the 19th century would see far stronger ties between transportation and oil, culminating in transformations that would become fully visible only once large oil strikes and more advanced drilling arrived. But even without those bigger developments, the mid-century evidence was clear: oil had begun inching its way into the beating heart of 19th-century transport, step by careful step.

CHAPTER 16: OIL AND POLITICAL POWER BEFORE THE LATE 19TH CENTURY

Introduction

Long before oil became a centerpiece of modern geopolitics in the 20th century, local rulers, entrepreneurs, and colonial powers in the 19th century had begun to grasp that controlling petroleum resources conferred certain advantages. While the scale of the oil trade was modest compared to later eras, competition for bitumen pits, seepages, and shallow wells foreshadowed the more extensive struggles that would follow. In many cases, these early conflicts or negotiations unfolded quietly, with limited documentation, yet they paved the way for the legal and political structures around resource ownership that characterize modern times.

In this chapter, we will examine how oil influenced political power well before the late 19th century. We will see how local dynasties or governments recognized that bitumen or oil seepages could supply crucial materials—whether for trade, domestic development, or limited military applications. We will also explore how colonial authorities, indigenous leaders, and merchants jockeyed for control of known deposits, occasionally sparking small disputes or forging alliances. Though overshadowed by more profitable commodities such as gold or spices, petroleum's strategic value was already emerging, and its role in shaping political relationships was more significant than many contemporaries realized.

1. The Broader Context of Resource Competition

1.1. 19th-Century Imperial Ambitions

Throughout the 1800s, European powers such as Britain, France, and Russia expanded their colonial empires. They sought natural resources—precious metals, fertile lands, and strategic ports. Petroleum did not yet top their priority lists, but it appeared often enough, especially in regions like the Middle East, the Caucasus, and parts of Southeast Asia. A savvy colonial governor might note the presence of bitumen seeps in a territory, if only as a minor supplement to local revenues or as a useful commodity for public works.

Colonial administrators approached oil as a curiosity or a small economic asset, but they did not ignore it. Some insisted that local populations gather and trade it through official channels, ensuring the crown or empire earned a modest tax or export duty. Others recognized that securing bituminous deposits could benefit naval shipyards, providing waterproofing materials or marine lubricants.

1.2. Local Rulers and the Desire for Economic Leverage

In regions not fully under colonial control, local dynasties, tribal chieftains, or regional strongmen often recognized that controlling seepages or tar pits gave them bargaining power. If neighboring territories needed pitch for boat repairs or lamps, the group in control could levy fees or shape trade routes to their advantage. These local power structures sometimes formed alliances with outside merchants, offering exclusive rights to gather oil in exchange for weapons, luxury goods, or political support.

These relationships did not usually dominate the headlines of 19th-century geopolitics. Far more attention was paid to disputes over gold mines, fertile plantations, or trade in commodities like cotton and tea. Still, in certain pockets of the world, petroleum subtly influenced how leaders exercised authority and carved out small economic niches.

2. Case Study: Baku and the Caucasus

2.1. Historical Background

Baku, on the western shore of the Caspian Sea in modern-day Azerbaijan, had a long heritage of surface oil collection. By the 1800s, it was part of a contested region where Persian, Russian, and Ottoman interests converged. The local khanates that once ruled Baku recognized the economic significance of the nearby oil fields, using them to bolster local treasuries. Later, as the Russian Empire expanded southward, officials in St. Petersburg saw potential in integrating Baku's oil into the empire's resource network.

2.2. Russian Expansion and Control

With the Treaty of Gulistan (1813) and subsequent agreements, Russia consolidated control over large parts of the Caucasus, including Baku. Early in the 19th century, Russian administrators began issuing licenses or concessions for shallow oil wells and pits. Though these operations were small compared to modern standards, they gave the Russian Empire a modest supply of oil for local use in southern territories.

Officials recognized the strategic dimension: secure oil meant better lubrication for machinery, local revenue from exports to neighboring regions, and potential materials for the imperial navy. They placed garrisons near key oil-producing villages, ensuring stability and discouraging brigands or rival claimants from seizing the pits. Over time, the region's output rose, creating a mini-boom in the local economy and attracting Russian settlers and entrepreneurs.

2.3. Local Resistance and Accommodation

Not everyone welcomed Russian oversight. Some indigenous landowners resented the empire's attempts to regulate or tax their ancient seepage rights, which had been privately held for generations. Periodic skirmishes flared, and local chiefs sometimes refused to recognize Russian drilling permits. The Russian military presence, combined with incentives for collaboration, eventually forced most local elites into a cooperative stance.

Still, certain families managed to strike deals that let them retain partial control or at least earn royalties on extracted oil. This arrangement showed how local authority figures adapted to the new power structure. By the 1840s and 1850s, Baku's oil production was a stable, if relatively small, part of the empire's resource portfolio, quietly foreshadowing the region's explosive growth later in the century.

3. The Ottoman Empire, Persia, and Middle Eastern Bitumen

3.1. Ottoman Administration of Natural Resources

In the Ottoman Empire, resource management varied from province to province. Some governors oversaw areas rich in bitumen seeps or asphalt lakes, using them to supply local or naval shipyards. While large-scale drilling was non-existent, the empire recognized that bitumen had tactical uses for fortress repair, boat caulking, and possibly small incendiary devices in warfare.

However, the empire's central bureaucracy rarely devoted special attention to these deposits unless a local crisis prompted it—such as a revolt in a region controlling a key resource. Even then, the demands of maintaining a sprawling empire meant that petroleum remained a footnote, overshadowed by grain supplies, taxation from agriculture, and controlling trade routes. Nonetheless, local pashas or beys occasionally leveraged access to bitumen or oil seepages to negotiate with the central government or with foreign merchants seeking specialized materials.

3.2. Persian Influence and Tribal Dynamics

Persia (Qajar Iran in the 19th century) also had territories known for natural seepages. Tribal leaders in border regions might trade lumps of asphalt or small quantities of oil to passing caravans. In certain mountainous districts, local clans recognized the strategic advantage of controlling passable routes near seepages. Although the monarchy in Tehran had nominal authority, tribal autonomy made direct governance patchy.

As a result, local notables sometimes brokered deals with traveling merchants from Russia, Central Asia, or the Ottoman lands, bartering bitumen for weapons or essential goods. The Qajar court, stretched thin by internal rivalries and external pressures from Russia and Britain, had limited capacity to enforce a centralized oil policy. Still, it recognized that controlling key resources could help maintain a semblance of stability and revenue. In minor but telling ways, these power plays over bituminous deposits highlighted how resource politics operated at both imperial and tribal levels.

4. North American Context: Early Territorial and State Governments

4.1. The United States and Resource Policies

In the early 19th century, the young United States government focused primarily on expanding its frontier, negotiating with Native American nations, and establishing states. Petroleum was not a national priority, but certain states—like Pennsylvania, New York, and Kentucky—passed basic laws about resource ownership. Generally, the prevailing principle was that landowners held the rights to the minerals (including oil) beneath their property, though this concept was still evolving.

Small disputes occasionally arose when multiple parties claimed seepages along property lines or near shared waterways. County courts were forced to interpret old property deeds in the context of new resource definitions. Some local judges recognized that "rock oil" had economic potential, so they ruled in ways that clarified or extended property rights to encourage modest investment. These early legal precedents, though rarely spotlighted, formed a basis for more robust mineral-rights frameworks down the line.

4.2. Indigenous Nations and Resource Sovereignty

Native American nations in regions with seepages (like parts of Pennsylvania or the Ohio Valley) had their own customs around resource usage. They might have viewed oil as a communal good for medicinal or ceremonial purposes. As American settlers encroached, friction arose over who legitimately controlled these seeps. Some tribes tried to defend traditional usage, while land-hungry speculators purchased large tracts—sometimes dishonestly, sometimes via treaties that indigenous leaders felt compelled to sign under duress.

Although oil was not the primary driver of American-Indian conflicts in the early 1800s (land for farming, timber, and settlement overshadowed everything else), it occasionally featured in negotiations or local tensions. The seeds of future disputes over mineral wealth were sown as tribal nations recognized that Americans placed commercial value on resources that had once been purely for local communal benefit. In a few documented instances, Native Americans bartered the "healing waters" or small jars of crude to settlers, unwittingly contributing to the latter's belief in potential fortunes beneath the land.

5. Local Strongmen, Merchant Alliances, and "Resource Militias"

5.1. Informal Militias Protecting Seepages

In certain rugged or remote territories—whether in parts of Central Asia, mountainous Middle Eastern districts, or the backcountry of the Americas—local warlords or strongmen formed small militias to guard bitumen pits or known oil seepages. They might charge caravans or traveling merchants a fee to fill containers, ensuring a steady trickle of revenue. Because these areas were beyond tight imperial or colonial oversight, the militia leaders effectively controlled resource access through force or threats.

For the local population, this arrangement could be both blessing and curse. On the one hand, the militia's presence deterred outside bandits or rival groups from raiding the seepages. On the other, villagers had to accept the warlord's authority, pay taxes in kind or labor, and forfeit any illusions of independent trading. The militia heads sometimes formed uneasy alliances with nearby officials or even cross-border merchants, so long as the arrangement was profitable.

5.2. Merchant Cartels and Oil-for-Salt Trades

In a few border zones, merchants formed cartels, pooling resources to control the distribution of bitumen or oil from local seeps. They might coordinate prices and volumes, shipping barrels to towns needing tar for roof repairs or boat building. In return, they imported salt, textiles, or basic manufactured goods for local consumption. Because these trades were small and often overshadowed by more lucrative deals (like spices or metals), official records rarely highlighted them.

Yet these merchant alliances wielded enough influence in some communities that local chieftains or small colonial garrisons recognized them as de facto authorities on resource matters. The cartels negotiated protective escorts, bribed officials to secure transport passes, and occasionally financed well-digging if they believed it would yield more product. These early forms of "oil commerce" demonstrated that, even on a tiny scale, the interplay of resource control and commerce could shift local power balances.

6. Maritime Powers and Pitch Dependencies

6.1. Naval Significance of Tar and Bitumen

Although the age of wooden sailing fleets was gradually giving way to steam-powered ironclads, most 19th-century naval vessels still featured significant wooden components—hulls, deck planks, masts, and rigging. These surfaces needed sealing and waterproofing, traditionally provided by pine tar or pitch. However, in regions lacking extensive pine forests, petroleum-derived bitumen offered an alternative.

Navies like Britain's Royal Navy or the French and Spanish fleets typically relied on well-established tar imports, but in certain colonial outposts, local bitumen from surface deposits supplemented pine resin shipments. Admiralties thus had a mild interest in any territory rumored to hold such resources. If a colony's governor reported large bitumen pits, he might receive instructions to evaluate their quality, potential yield, and reliability as a supply source for minor naval repairs.

6.2. Rivalries in Coastal Territories

When multiple powers vied for influence in coastal zones or island colonies, controlling bitumen or tar resources could offer a strategic edge, albeit a small one. For example, in the Caribbean, some islands contained natural pitch lakes or seepages that offered easily gathered lumps of tar for repairs to ships. Rival colonial governors angled for treaties with local tribes or introduced taxes to ensure they, rather than a competing empire, reaped the resource's benefits.

These rivalries rarely escalated into full-blown wars over oil. More significant motivations—sugar plantations, gold shipments, or strategic harbors—dominated international competition. Still, bitumen control sometimes played a supporting role in negotiations or local disputes, reinforcing the notion that even lesser-valued resources could sway minor power dynamics. Over time, as steam-driven iron warships emerged, the direct naval reliance on wooden pitch sealing declined, but by then petroleum's other uses, like lubrication, were becoming more apparent.

7. Diplomatic Maneuvers and Oil Concessions

7.1. Early Concessions in Eastern Europe

In places like Galicia, foreign investors occasionally sought concessions from the Austro-Hungarian authorities. While the empire valued major mining sectors (salt, coal, metals), officials recognized that small petroleum deposits could provide local lamp oil and some export potential. They granted short-term monopolies or lease rights to entrepreneurs who promised to invest in well-digging or refining.

These contracts often contained clauses about taxation, the number of local workers employed, and minimal production targets to avoid speculation without development. While mostly small-scale, these deals planted the seeds of future concession-based approaches—where a governing body licenses resource extraction to a private entity in exchange for royalties, fees, or infrastructure improvements.

7.2. British and French Expeditions in the Middle East

British and French explorers who ventured into the Middle East sometimes stumbled upon surface seepages that local inhabitants had used for centuries. Their dispatches back home mentioned potential for future exploitation, though immediate action was uncommon. British East India Company officials, for instance, had bigger priorities like controlling trade routes and ensuring a steady flow of cotton, tea, or opium.

However, a handful of forward-thinking officials quietly advocated for securing rights around promising bitumen deposits. They reasoned that in the event of conflict, these resources might prove useful for naval logistics. Others believed that controlling the local trade in pitch could yield modest revenue or strengthen Britain's or France's foothold in contested regions. Though overshadowed by grander imperial strategies, these tidbits of interest foreshadowed the eventual scramble for Middle Eastern oil in the 20th century.

8. Oil in Military Contexts Before the Late 19th Century

8.1. Limited Use of Incendiaries

Historically, flammable materials like "Greek fire" or naphtha had played roles in medieval warfare. In the 19th century, advanced firearms and artillery overshadowed such tactics, but militaries occasionally considered or tested oil-based incendiaries for siege or naval defense. Some armies used thick petroleum or bitumen mixtures in rudimentary bombs or to create defensive barriers.

The effect was minor, as gunpowder-based weaponry was far more practical. Nonetheless, certain fortress commanders in the Middle East or Eastern Europe recognized that a supply of oil could bolster perimeter defenses in a desperate situation. They might store barrels of tar or bitumen in strategic points, ready to pour onto attackers or set ablaze if walls were breached. While rarely decisive in major conflicts, these old traditions of "fire weapons" persisted on a small scale.

8.2. Maintenance of War Machines

A more common military use of oil was for lubrication of cannons, artillery carriages, and other mechanical devices. Armies also had growing fleets of steam-driven wagons or small railway lines supporting campaigns. These required grease that wouldn't fail in hot, dusty conditions—an environment where refined petroleum lubricants sometimes performed better than tallow. Quartermasters who recognized these benefits might secure shipments of oil from local sources or import them from known refining hubs.

Armies seldom formalized large petroleum procurement strategies at this stage. But each successful application—for example, an artillery unit praising how a new oil-based grease reduced barrel rust—reinforced the resource's value. This subtle acceptance in military logistics showed that petroleum had strategic dimensions beyond mere commercial trade, albeit not yet on a grand scale.

9. Social Perceptions of Oil as a "Minor Strategic Resource"

9.1. Newspapers and Public Narratives

19th-century newspapers generally focused on politics, wars, industrial developments, and major commodities. Mentions of oil as a strategic asset were fleeting, typically confined to local reports about new seep discoveries or unusual uses in shipping. Occasionally, a small editorial might speculate that controlling local tar pits gave a city or region an edge in sealing roads or constructing better fortifications, but these pieces were overshadowed by more dramatic events.

In some corners of society, rumors swirled about the hidden riches of "rock oil," though these rumors often conflated it with fantasies of gold or gemstone deposits. The ordinary person, if they heard about oil in a political context, might assume it was simply another minor resource in an empire's vast domain, not something that could tip the scales of power.

9.2. Cultural Attitudes in Resource-Rich Areas

Communities living near active seepages or wells had a more direct understanding of oil's significance. They saw how local elites or colonial administrators taxed production, or how traveling merchants paid premiums for lumps of pitch. They might also witness occasional conflicts over well rights or the presence of militias guarding a valuable deposit. In such locales, oil was not a remote curiosity but a material shaping everyday economic and political realities.

Over time, these cultural attitudes reinforced a sense that if external authorities—imperial officials, local warlords, or merchant cartels—wanted to control the resource, it must have genuine value. This recognition set the stage for more intense political maneuvering once drilling technology advanced and larger yields became feasible.

10. Challenges to Centralized Control

10.1. Difficulties in Regulating Dispersed Seeps

One reason oil did not dominate 19th-century political power structures was the difficulty of controlling scattered seepages. Unlike a single diamond mine or gold deposit, petroleum often emerged in multiple small pockets spread over wide territories. Governments lacked the bureaucratic machinery to monitor each seep or pit, and local entrepreneurs could quietly gather oil without official oversight.

This decentralization meant that no single entity could easily monopolize supply—unless, like in Baku, the region had unusually high concentrations of seepages. Thus, even if a ruler or colonial administrator recognized oil's value, imposing strict monopolies was logistically challenging. The resource's scattered nature ironically prevented it from becoming the focus of large-scale political intrigue on par with, say, the spice trade or silver mines.

10.2. Technology Gaps and Small Extraction Scales

The limited technology for drilling also meant that even known deposits could not be fully exploited. Most operations produced only modest barrels per day, insufficient to catalyze major political or economic revolutions. Indeed, controlling a thousand barrels of oil annually might enrich a local baron or landowner, but it did not alter global power balances.

Without large outputs, there was little impetus for major powers to invest heavily in secure transport networks, advanced drilling rigs, or military deployment. Instead, resource control remained a micro-level phenomenon—important within its locale but rarely impacting broader imperial strategies. This remained the norm until more advanced drilling, refining, and distribution methods emerged, mainly in the latter half of the century.

11. Foreshadowing Future Strategic Importance

11.1. Visionary Observers and Publications

A minority of 19th-century writers speculated that if drilling improved and the demand for oil-based fuels or lubricants grew, controlling the best deposits could become a significant political advantage. Some geologists or travelers penned articles in scientific journals hinting at the possibility of large-scale extraction in places like Pennsylvania, Baku, or the Middle East.

While these observers were rarely heeded by mainstream policy-makers, their writings left an intellectual breadcrumb trail. Future industrialists and politicians, in the latter part of the century, would revisit these earlier texts to argue for deeper wells, dedicated infrastructure, and stronger state involvement. In that sense, the seeds of the "oil question" were sown well before public consciousness caught up.

11.2. Subtle Shifts in Treaties and Governance

In a few treaties, land sales, or colonial charters, references to "minerals and bitumen deposits" appeared. Diplomatic negotiators occasionally included language ensuring that if one power discovered profitable oil resources, it had to share or pay royalties to the other. These clauses, though overshadowed by broader negotiations, hinted that savvy diplomats understood oil might not remain minor forever.

Additionally, certain regional authorities began standardizing policies for well permits or seep usage. They introduced early forms of taxation or licensing specifically referencing "oil wells," distinct from those for salt or metals. Though typically small-scale, these legislative efforts signaled a nascent awareness that oil needed its own regulatory approach—anticipating the robust petroleum laws of the 20th century.

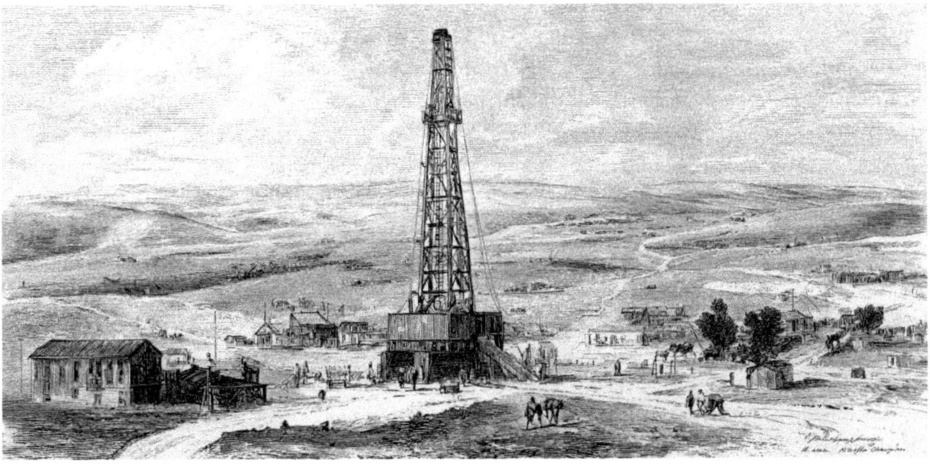

CHAPTER 17: CULTURAL AND SOCIAL IMPACT OF OIL IN THE 19TH CENTURY

Introduction

Throughout the 19th century, oil slowly but steadily worked its way into diverse aspects of daily life. While previous chapters explored drilling techniques, refining, and the political significance of oil, this chapter takes a different approach. It examines how oil shaped and reflected cultural norms, social structures, and communal identities. From small rural settlements near seepages to growing industrial towns using oil-based products, people came to view petroleum in many ways: as a potential miracle cure, a source of local pride, a step toward progress, or even an object of suspicion.

In these pages, we look at how newspapers reported on new oil discoveries, how religious or medicinal beliefs featured oil as a special substance, how it influenced class distinctions in certain regions, and how it played into evolving consumer behaviors. We also see how community events, folk stories, local charities, and even protest movements took shape around the presence or absence of "rock oil." Although overshadowed by more dominant 19th-century social forces—like rising nationalism, the industrial revolution, and sweeping reform movements—oil still had a noticeable cultural footprint. By following these stories, we gain a broader understanding of petroleum's significance in people's daily experiences, long before it became the globe-spanning industry we recognize today.

1. Newspapers, Broadsheets, and the Public Perception of Oil

1.1. Early Press Coverage of "Rock Oil" Discoveries

In the early and mid-19th century, local newspapers and broadsheets often served as the main source of information for communities. When news reached editors that someone had uncovered a new seep or drilled a shallow well that yielded promising amounts of oil, the story might headline the local paper—especially if the region had few other major events to discuss. Journalists typically focused on the novelty, calling it "strange black liquid," "rock oil," or "Seneca oil" in certain parts of North America.

Some articles relayed wild speculations: claims that the oil could cure dozens of ailments, create new fortunes overnight, or even end the reliance on whale oil if production expanded. Other writers took a more cautious tone, reminding readers that many wells produced limited output and that refining processes remained rudimentary. Across the board, curiosity abounded. People who read these reports developed a sense that oil might hold secrets still to be uncovered.

1.2. Sensational Stories and Penny Press Papers

The penny press phenomenon—cheap newspapers aimed at mass audiences—emerged in some cities by the 1830s and 1840s. These papers thrived on sensational or entertaining stories to boost sales. Oil discoveries fit neatly into that niche: a mysterious substance rising from the ground, with possible links to health miracles or quick fortunes, made for exciting reading. Some articles stretched facts or cited dubious sources, claiming spectacular "gushers" or the discovery of "liquid gold." While many readers took such tales with a grain of salt, the repeated exposure helped keep oil in the public eye.

People in distant towns who might never see a real seep or drilling rig still formed opinions about oil from such sensational news. For them, petroleum became almost legendary—a resource that might solve problems, open new opportunities, or lead to ruin if mismanaged. In this way, newspapers shaped broad social narratives around oil, influencing whether communities welcomed or feared new drilling ventures.

1.3. More Serious Journals and Scientific Debates

On a more serious note, scientific journals and technical publications—though read by fewer people—offered detailed accounts of drilling methods, refining experiments, and the chemical makeup of oil. Articles published in early geology or engineering journals sometimes contradicted the hype found in penny press stories. They stressed measured observations, caution about overestimating well yields, and the challenges of refining.

This dual press environment meant that educated elites or specialized professionals often had a more tempered view of oil's prospects, while the general populace might see it as a dramatic novelty. The gap between these perspectives could fuel local debates: a geologist's cautious stance might clash with a merchant's ambitions, or a small farmer's excitement might outpace the realities explained by an academic chemist. Ultimately, these contrasting media narratives shaped the early social climate in which oil's cultural significance took root.

2. Medical Beliefs, Miracle Cures, and Patent Medicines

2.1. Folk Medicine Traditions

Many communities living near seepages, be they in North America, Eastern Europe, or parts of Asia, long held folk traditions that attributed healing properties to crude oil.

People might apply thick bitumen to skin sores or rub lighter fractions on aching joints. Indigenous groups in North America used "burning springs" or "healing waters" in ceremonies for centuries. Such practices often influenced European settlers or travelers, who saw local inhabitants claim relief from various ailments.

Though few formal studies backed these claims, word of mouth was powerful. Tales of a neighbor's wound healing faster with an oil-based salve circulated through close-knit towns. These stories contributed to the notion that oil contained special properties beyond mundane uses like lighting or lubrication. Even if actual medical benefits varied, the belief in petroleum's healing potential persisted and spread.

2.2. Rise of Patent Medicines

In the 19th century, patent medicines flourished—often unregulated concoctions sold under grand names, promising to cure everything from headaches to kidney disease. Some of these potions included forms of crude or partially refined oil. Advertisements in newspapers featured glowing testimonials from "satisfied customers," while traveling salesmen set up shop in public squares to pitch their wares.

Products labeled "Seneca Oil" or "American Medicinal Oil" sometimes found eager buyers. While the exact recipes were secret, many presumably included a fraction of petroleum mixed with herbs, alcohol, or other additives. Because modern pharmaceuticals and scientific oversight were limited, consumers relied on anecdotal evidence and local rumor. Some might indeed feel temporary relief, perhaps due to the warming effect of an oil-based rub or mere placebo. Meanwhile, unscrupulous sellers used these success stories to inflate claims, fueling the impression that crude oil was a cure-all.

2.3. Religious and Ritual Contexts

In certain religious communities, items like "holy oil" or "sacred ointments" played roles in ceremonies. Though not always derived from crude petroleum, there were occasional crossovers, especially in areas where local clergy might adopt folk beliefs. Some small sects integrated bituminous substances into healing rites or blessings, attributing spiritual significance to the thick, dark fluid seeping from the earth.

These practices varied widely by region, but they highlight how oil's cultural meaning went beyond economics. For believers, it was not just a resource but a mystical sign, possibly tied to blessings from higher powers. This spiritual dimension could create friction with more secular or scientific viewpoints, as devout followers insisted that oil-based cures worked through divine will, while doctors or chemists looked for rational explanations.

3. Class Distinctions and Oil's Role in Society

3.1. Access to Refined Oil for Lighting

By the mid-19th century, in places where refining had advanced, a portion of the population had access to petroleum-based lamp fuels. Because these products often remained relatively expensive compared to cheaper but smokier alternatives, they became something of a status symbol in certain circles. Wealthier families who disliked the smell of whale oil or tallow candles might pay a premium for a clearer-burning oil lamp, appreciating the brighter flame and reduced soot.

Though still overshadowed by other fuels, these refined oils signaled a household's modern outlook and disposable income. In industrializing cities, a merchant or factory owner might display this new lamp proudly, showing visitors that they kept abreast of the latest technological trends. Meanwhile, working-class or rural families generally stuck with homemade candles or simpler fats, perceiving refined petroleum as a luxury product beyond their means or local supply.

3.2. Employment and Emerging Oil Communities

Wherever modest oil booms occurred—such as the small-scale expansions in Galicia, Baku, or certain North American towns—new job opportunities arose. Local men found work digging pits, carrying oil, or operating basic refining stills. Craftsmen specialized in building barrels or rigging simple mechanical pumps. Over time, these tasks formed a minor sub-economy within the community, creating divides between those who thrived from oil-related work and those who continued in farming or other trades.

In some villages, well owners or small-time speculators became local elites, earning enough money from oil sales to build larger homes or invest in land. Their success stories fueled envy or emulation among neighbors, shaping the social fabric with new wealth-based hierarchies. But the volatility of well yields and the uncertain market for crude or refined products also meant fortunes could be lost as quickly as gained.

3.3. Philanthropy and Civic Projects

A handful of more prosperous oil entrepreneurs supported community causes, donating to local charities, schools, or churches. Though rarely on the scale of 20th-century oil philanthropists, these earlier benefactors built small libraries, funded street improvements, or contributed to poor relief—particularly in smaller North American or European towns experiencing mild oil-related growth. Newspapers might praise these gestures, reinforcing the notion that oil money could uplift local society.

Not all gave altruistically. Some donors aimed to polish their reputations or secure influence in municipal decision-making. Yet even these pragmatic motives underscored oil's evolving role: once simply a messy black goo, petroleum now provided a path to social prestige and civic leadership. The community's perception of an oil businessman or landowner might shift from suspect gambler to respected patron if he funneled profits into visible, beneficial projects.

4. Fairs, Exhibitions, and Public Demonstrations

4.1. Local Festivals Showcasing New Fuels

It was not uncommon in the 19th century for towns to hold agricultural fairs or mechanical exhibitions, where inventors and merchants displayed their innovations. At these gatherings, traveling oil salesmen might set up a stall to demonstrate how refined petroleum lamp fuel burned cleaner than typical tallow candles. They might also show small lumps of tar that could repair leaky roofs or waterproof barrels. Crowds, intrigued by the novelty, asked questions or purchased small quantities to try at home.

Such fairs also featured side attractions—perhaps a demonstration of an "oil-based healing remedy" or a curious mechanical device lubricated with petroleum. These displays fed the public imagination, merging entertainment with practical education. People who had never seen raw crude might handle a sample, smell its pungency, and watch it burn. As a result, new visitors left with a heightened awareness of oil's possibilities, sparking dinner-table discussions about the pros and cons of adopting it in everyday life.

4.2. National and International Exhibitions

On a larger stage, mid-19th-century world's fairs, such as London's Great Exhibition of 1851 or subsequent industrial expositions, sometimes included small oil-related showcases. While overshadowed by major attractions—like colossal machinery, new textiles, or architectural marvels—oil refining demonstrations and display stands featuring new lamp models still drew interest. Attendees saw chemical apparatuses that distilled crude, read about "remarkable wells" in distant places, or peered at samples of different petroleum fractions.

These international exhibitions were crucial in broadcasting the existence of oil to broader audiences. Diplomats, investors, engineers, and curious members of the public mingled and learned from each other. They might pick up brochures, share contacts, or pass tips along to acquaintances back home. Although oil's presence at these fairs remained comparatively small, it nudged the resource's reputation from local oddity toward something with genuine global relevance.

5. Popular Literature, Travelogues, and Oil's Romantic Image

5.1. Romanticized Travel Writing

The 19th century saw an explosion of travel literature, as explorers and adventurers recounted journeys into remote regions. Writers who stumbled upon oil fields—like those in Baku or along the Caspian, or seepages in North America—often added dramatic flourishes to their descriptions. They portrayed black pools glimmering under the sun, mysterious natural flames erupting from the ground, or local communities worshipping or fearing the "fire wells." These romantic narratives captured the public's imagination.

Readers back in urban salons or quiet villages devoured such accounts, associating oil with exotic locales and wonder. While geologists or seasoned merchants might roll their eyes at the hyperbole, these stories strengthened the notion that petroleum was tied to hidden riches, curious rituals, and untapped power. Over time, some adventurous souls ventured to these areas, hoping to witness the marvels or even profit from them, driven partly by the aura that travelogues cultivated.

5.2. Fiction and Serialized Stories

Occasionally, fictional works—serialized in magazines or sold as cheap booklets—incorporated oil as a plot element. A character might discover a secret seep on family land, leading to family feuds, suspicious accidents, or comedic misunderstandings. Writers used oil's mystique as a device to explore greed, ambition, or rivalry, especially in smaller communities where a minor resource find could upend social norms.

Such tales sometimes overlapped with moral lessons: a cunning speculator who overstates a well's capacity might fall into disgrace, or a humble family that prudently manages a seep might gain moderate prosperity. While not mainstream literary themes, these narratives reached enough readers to shape broader cultural attitudes—portraying oil as both a potential blessing and a cause of social strife, depending on one's moral compass and approach.

6. Superstitions, Fears, and Local Reactions

6.1. Fears of "Unholy Substances"

In conservative or deeply religious communities, the dark color, pungent smell, and flammable nature of oil could evoke suspicions. Some believed that substances oozing from the earth signaled demonic or cursed origins. Rumors circulated of families who used crude oil for lamps suffering ill fortune or ill health, leading neighbors to blame the "unholy substance."

Furthermore, drilling practices that brought the "subterranean liquids" to the surface might spark local anxiety: were these unnatural intrusions into God's domain? Did it risk releasing harmful vapors or angering spirits? A well driller in such an environment could face ostracism unless community leaders were convinced of the well's practical benefits or moral neutrality.

6.2. Accidents and Shifting Public Opinion

Small blowouts, fires, or spills occasionally occurred in early drilling or refining sites. When these accidents destroyed property or led to injuries, local rumor mills sometimes pinned blame on the risky nature of oil itself. Families that lost crops to an oil-tainted stream might lead protests against further drilling, warning neighbors that the resource brought more harm than good. Others, witnessing the economic advantage gained by well owners, defended continued extraction.

These localized controversies influenced how entire towns viewed oil going forward. A single tragic fire might overshadow the resource's benefits, souring community relations for years. Conversely, if a crisis was managed well—perhaps with the well driller compensating affected families—people might come to see such mishaps as normal teething pains in an evolving industry. Overall, local experiences with mishaps played a major role in shaping cultural acceptance or rejection of petroleum.

7. Social Conflict and Worker Protests

7.1. Labor Tensions in Emerging Oil Fields

Wherever moderate oil activity clustered, such as in parts of North America or Eastern Europe, a small labor force emerged to dig or drill wells, transport barrels, or run primitive refineries. Workers often earned low wages, performed dangerous tasks, and had minimal job security if a well ran dry. They might also be subject to health hazards like inhaling oil fumes or handling flammable materials without safety gear.

Protests occasionally arose when wages were cut or working conditions worsened. Laborers demanded fair pay or protective measures, though formal labor unions remained rare in early 19th-century oil communities. These small-scale strikes, or walkouts, sometimes disrupted local production temporarily. Employers might respond with threats of dismissal, or they might make minimal concessions to keep operations running. In either case, the presence of such conflicts demonstrates that even small petroleum ventures could replicate the broader industrial era's labor tensions.

7.2. Community Divisions over Drilling Expansion

When operators sought to expand drilling, communities could become polarized. Some farmers hoped to lease their land for wells, anticipating royalty payments or direct profits. Others feared contamination of water sources, reduced farmland quality, or the influx of transient workers who might disrupt local norms. Meetings in town halls grew heated, with each side lobbying local officials for permits or bans.

In some instances, religious leaders or prominent citizens stepped in to mediate. They might organize committees to inspect proposed drilling sites or consult local experts. The outcomes varied: a few towns imposed strict regulations, limiting well depth or requiring certain safety measures, while others embraced drilling wholeheartedly, hoping for an economic windfall. These debates laid early groundwork for the later rise of more formal zoning laws and resource regulations.

8. Adaptation in Household and Domestic Life

8.1. Lamps, Stoves, and Household Innovations

In the 19th century, the typical home lighting system depended on candles or whale oil lamps. As refined petroleum fuels became slightly more available, some households replaced or supplemented their older lamps with oil-burning versions. Women who managed domestic tasks appreciated the brighter, more stable light for evening sewing or reading, though they often disliked the lingering odors if the oil was not well-refined.

A few inventors also tinkered with small petroleum stoves, hoping to create a portable cooking device. These contraptions, while never mainstream in mid-century, found limited use among travelers or in remote outposts where wood or coal was scarce. Though not as widely adopted as later kerosene stoves, they signaled a willingness among households to experiment with new petroleum-based conveniences, provided they seemed safe and practical.

8.2. Household Remedies and Cleaning Agents

Beyond lighting or heating, some families used petroleum-based solutions for cleaning or pest control, especially in areas with consistent supply. A heavier oil fraction might remove stubborn stains or grease squeaky hinges. Others believed that a petroleum-based wash repelled insects from barn doors or stable walls. Many such remedies circulated in local homemaking guides or oral tradition, forming a minor subculture of "petroleum hacks."

However, these uses remained overshadowed by simpler, established products like lye soap or vinegar solutions. People who did adopt oil-based cleaners had to be vigilant about fumes or accidental ignition. Even a small lamp flame in the vicinity could spark a dangerous fire if the cleaning oil was mishandled. Despite these risks, the variety of domestic experiments with petroleum underscores its creeping presence in the 19th-century household sphere.

9. Charitable Foundations, Mission Work, and Oil Patronage

9.1. Oil-Funded Local Charities

Though large philanthropic organizations funded by oil wealth did not emerge until later eras, some smaller charitable endeavors in the 19th century traced their resources to local oil entrepreneurs. A well owner who made moderate profits might donate a portion to the local church, which then supported an orphanage or a basic medical clinic. In a few known cases, traveling missionaries in Eastern Europe or the Caucasus received donations from families who gained wealth from shallow wells, helping them build chapels or schools.

These acts rarely garnered wide publicity, as the amounts were small by global standards. Nonetheless, they reflected an early pattern: philanthropic gestures tied to petroleum, reinforcing the idea that resource extraction could aid community welfare. In turn, local religious leaders might publicly praise the donors, conferring social approval and further embedding oil in the moral and cultural landscape.

9.2. Missionaries Introducing Oil Products Abroad

Conversely, missionaries traveling from Europe or North America to distant lands sometimes carried lamp oil or small refining kits, hoping to demonstrate "modern methods" to the populations they served. They believed such practical tools could help them gain trust and illustrate the benefits of Western technology, thereby aiding their missionary goals. In some African or Asian communities, these foreign missionaries introduced petroleum lamps for nighttime reading in mission schools, sparking local curiosity about the substance.

Although this exchange was limited, it hints at a broader cultural interplay: missionary zeal, technology transfer, and the infiltration of oil-based goods into corners of the world where local inhabitants might otherwise have remained unaware of them. Mission reports, in turn, might describe how well these lamps functioned or how local communities reacted, feeding back into the global conversation about oil's diverse impacts.

10. Shifts in Consumer Behavior and the Road to Mass Adoption

10.1. Emerging "Consumer Culture"

The 19th century witnessed the slow birth of what we might call a consumer culture, driven by industrialization, rising incomes (in certain segments), and expanding retail networks. Although petroleum-based products constituted a sliver of the consumer market, their appearance in general stores or traveling peddlers' wagons signaled that everyday people had more choices in lighting, home remedies, and lubricants.

As brand awareness grew—albeit modestly, with some refined oils marketed under specific names—customers started distinguishing between high-quality and low-quality oils. Word of mouth and limited advertising taught them that poorly refined petroleum might smoke or smell badly, whereas a "double-distilled" brand offered better performance. This budding consumer discernment paved the way for stronger, more recognizable oil brands in the future, tying into the rise of formal marketing strategies.

10.2. The Power of Demonstrations and Testimonials

A key driver of consumer acceptance was personal recommendation. If a neighbor saw that the new lamp oil burned steadily all evening without blackening the chimney, they might buy a small bottle next payday. If a relative's knee pain seemed alleviated by a petroleum salve, another family might try it too. In close-knit rural settings, such direct peer influence often trumped official announcements or newspaper ads.

Town merchants, aware of this dynamic, sometimes hosted in-store demonstrations. They might invite customers to watch two lamps side by side—one with older whale oil, the other with refined petroleum—and compare brightness and smell. Observing the difference could persuade hesitant buyers. Thus, the social spread of petroleum usage was often a bottom-up phenomenon, reliant on neighbors, friends, or local shops showcasing the benefits in tangible ways.

CHAPTER 18: ENVIRONMENTAL VIEWS AND EARLY CONCERNS

Introduction

Modern readers often associate oil with large-scale environmental controversies—offshore spills, air pollution, and debates over climate change. However, even in the 19th century, smaller communities and local observers recognized that extracting and using petroleum could degrade land, contaminate water, and harm public health. Although formal "environmentalism" had not yet coalesced as a major social movement, scattered voices raised warnings about unregulated drilling, refining waste, and the messy aftermath of open-pit oil collection.

This chapter delves into how 19th-century societies perceived and managed the environmental impacts of oil. We will see how repeated spills near shallow wells led farmers to protest, how local authorities imposed early regulations on waste disposal, and how certain civic leaders or health officials worried that fumes or polluted waterways endangered communities. While overshadowed by the era's focus on industrial expansion, these early critiques and modest protections laid the groundwork for more sophisticated environmental debates in the following century. Understanding these roots highlights that environmental awareness, though limited in scope, did not arise overnight—it emerged gradually, in tandem with oil's growing presence in daily life.

1. The State of Environmental Awareness in the 19th Century

1.1. Pre-Industrial Views of Nature

For much of history, nature was seen as boundless, and the idea of polluting it to a harmful degree seemed less urgent in sparsely populated regions. Yet, industrialization in the 19th century began to change that. Cities grappling with coal smoke or factory runoff realized that large-scale production had consequences. Even so, many governments and communities lacked formal frameworks for managing industrial waste, including oil byproducts.

People who worried about the environment tended to focus on visible, local problems: foul-smelling water, fish kills in rivers, or tainted wells. They did not conceive of global-scale pollution or greenhouse effects. Instead, their motivations were practical: avoid disease, keep farmland productive, or preserve water for livestock. Oil, with its strong odors and sticky residues, could violate these local standards of cleanliness, prompting scattered complaints or rudimentary controls.

1.2. Rise of Public Health Concerns

Some mid-19th-century reformers, influenced by developments in medical knowledge and sanitation ideas, tied environmental cleanliness to public health. They observed that towns with contaminated waterways suffered outbreaks of diseases like cholera or typhoid. While bacteria theories were still emerging, community leaders recognized that any foul substance in the water supply might pose health risks.

Thus, when oil seepages or drilling operations threatened a stream or well, local residents often framed the issue in health terms rather than abstract environmental ideals. They might argue that runoff from an oil pit could spoil drinking water, cause livestock illness, or lead to noxious air that harmed children's lungs. These concerns led to the earliest documented attempts at restricting how and where oil producers disposed of waste, ensuring at least minimal care for local ecosystems.

2. Local Water Pollution: Streams, Wells, and Drinking Supplies

2.1. Oil Residues in Rivers and Lakes

One of the most common environmental grievances in early oil communities revolved around water pollution. When shallow wells overflowed or pit diggers dumped waste oil on the ground, rain could wash residues into nearby streams or lakes. In fishing communities, locals complained that oily films scared away fish or coated them in foul-smelling layers, ruining a source of food and income.

Farmers reliant on small creeks for irrigation sometimes discovered their crops wilting if oil-laden water coated the soil. Others lamented that livestock refused to drink from a slick-surfaced trough. These tangible impacts provoked immediate backlash, with farmers forming ad hoc committees to petition local authorities or confronting the well owners directly. The friction often put oil entrepreneurs on the defensive—if they valued community goodwill, they had to mitigate such pollution or pay damages, typically through informal negotiations rather than court battles.

2.2. Contamination of Public Wells

In towns that relied on communal wells, an accidental spill or seep too close to the water table could ruin the supply for everyone. Residents forced to walk farther for clean water or boil the well water found their daily routines disrupted. These incidents sometimes led to strong community backlash against further drilling, especially if the operators seemed careless.

Local newspapers would headline stories like "Town Well Spoiled by Oil Waste," stoking anger among readers. Public meetings might discuss banning new wells within a certain radius of the communal supply. In some cases, the well driller offered compensation or promised better containment to quell the uproar. While these solutions were ad hoc, they represented a preliminary form of environmental accountability—though not always consistent or enforceable over time.

3. Emissions, Fumes, and Health Complaints

3.1. Refinery Vapors and Local Nuisances

As small refineries sprang up near oil production sites in places like Galicia, Baku, and parts of North America, neighbors began complaining of continuous smoke, pungent fumes, and drifting odors. Early refining involved boiling crude in open or poorly sealed stills, releasing hydrocarbon vapors into the air. Without sophisticated condensers or pollution controls, stench and haze could blanket nearby houses, turning daily life unpleasant.

Some residents worried about respiratory illnesses in children or the elderly. Though modern toxicology was still in its infancy, people recognized that strong chemical smells often induced headaches or nausea. Women who hung laundry outdoors might find clothes tainted with an oily scent. Farmers might claim it spooked their livestock or reduced milk yields. These local annoyances contributed to a growing sense that the petroleum industry carried hidden costs, beyond the visible gains of job creation or improved lighting.

3.2. Incidents of Poor Air Quality

Although coal-burning factories remained the prime culprit for 19th-century urban smog, refineries in certain districts added to the mix. In industrial neighborhoods, where multiple factories clustered, city dwellers faced a haze of blended emissions—coal smoke from boilers, chemical fumes from dye works, and oily vapors from refineries. Overcrowded housing meant many families lived close to these polluting sites, enduring higher rates of respiratory ailments.

Occasional letters to the editor in local papers demanded better regulation or relocation of the most noxious refineries. Some city councils responded by requiring that refineries be built on the outskirts, near rivers or industrial zones, hoping to reduce the direct impact on residential blocks. This spatial separation strategy became an early form of zoning, aiming to keep "dirty" industries away from dense populations. However, enforcement varied widely, and refineries often found ways to remain near cheap transport routes, continuing to affect local air quality.

4. Improvised Disposal Methods and Waste Mismanagement

4.1. Dumping in Open Pits

Before the concept of industrial waste management took shape, operators typically dumped leftover sludge or heavy residues into open pits. This was common near oil wells and refineries, where spent or tarry byproducts had no immediate market. Rain could cause these pits to overflow, seeping into nearby land or waterways. Children playing in the area risked stepping into sticky sludge, or farmers might inadvertently track it onto roads, creating mess and hazards for horse-drawn traffic.

While dumping was cheap and easy, it fueled many local grievances. Landowners downstream or downhill from such pits felt their property values plummeting. They might confront the pit owners directly, demand cover with soil or stone, or ask local officials to intervene. Here and there, a conscientious refiner might line the pit with clay to reduce leaks, though these efforts were primitive. The general absence of formal guidelines left each site to manage waste on its own terms, often resulting in patchy success and frequent pollution incidents.

4.2. Burning Residues

Another disposal tactic was to set surplus oil or tar ablaze. Some saw it as a quick fix—burn the waste and let the smoke disperse. But these open fires released thick, acrid fumes, sometimes drifting over entire neighborhoods. If poorly supervised, a burning pit could spread flames to nearby fields or wooden structures. Fire brigades in some towns grew frustrated at repeated calls to extinguish refinery waste fires that got out of hand, occasionally endangering entire settlements.

Reports from local newspapers recounted nights where the sky glowed from these "tar fires," with residents complaining of soot falling on rooftops or into gardens. Opponents argued that repeated burning threatened both human and animal health, urging local leaders to ban the practice or require safer incineration techniques. Although a few municipalities passed rudimentary ordinances, enforcement remained minimal, leaving the hazard largely unaddressed in many regions.

5. The Emergence of Minimal Regulations and Community Rules

5.1. Township and County Ordinances

Faced with tangible pollution, some 19th-century townships or county authorities introduced basic rules on well placement, waste disposal, and refinement processes. For instance, a local board might ban drilling within a certain distance of a public well or require that oil operators build earthen dikes around pits to prevent overflow. Others mandated a small fee or license, using the revenue to pay for inspections (though enforcement was often lax).

These measures were not universal. Some areas, eager for economic growth, resisted any regulation. Entrepreneurs lobbied officials, warning that strict rules would chase away investment. On the other hand, citizens living with oil's downsides pushed for stronger protections. This tension meant that regulations typically reflected local power dynamics: communities with influential farmers or civic leaders who opposed pollution might enact stricter controls, while oil-boom towns courting outside capital might keep their codes lenient.

5.2. Private Contracts and Voluntary Guidelines

In places lacking formal governance structures, individuals sometimes negotiated private contracts or voluntary codes of conduct. A landowner leasing drilling rights might include clauses specifying that the driller must line waste pits, store refined oil in sealed barrels, or avoid dumping near streams. These deals resembled the forerunners of modern environmental compliance—though entirely private, enforceable only through the threat of lease cancellation.

Occasionally, local associations of well owners formed to maintain a collective reputation. They recognized that repeated pollution incidents could spark a public backlash or hamper their ability to sell oil. Thus, they agreed among themselves to follow certain disposal practices or share costs for building a communal waste pond. While not altruistic, such cooperation illustrated that some 19th-century operators grasped the financial and social incentives behind curbing environmental damage.

6. Early Scientific Voices on Environmental Risks

6.1. Geologists and Public Health Advocates

A small segment of geologists and medical reformers who studied emerging industries noted the hazards of unregulated oil extraction. In published papers or local lectures, they discussed how seepage into aquifers could taint wells or how thick residues might smother plant life in wetlands. Some also made connections between inhaling strong fumes and respiratory illness, urging better ventilation in refineries.

However, these warnings were rarely front-page news. Society valued scientific expertise, but the scientific community was smaller and lacked political clout to enforce wide-scale changes. Nevertheless, geologists who traveled to major oil-producing sites left behind detailed field reports describing environmental side effects—helping historians today see that such critiques existed, even if overshadowed by the push for industrial progress.

6.2. Local Physicians' Observations

Doctors serving rural or semi-industrial regions often encountered patients complaining of headaches, persistent coughs, or skin irritations linked to oil-laden surroundings. In their personal diaries or in letters to medical journals, they sometimes attributed these ailments to "miasmas" from stagnant oil pits or continuous smoke from refining. Though 19th-century medical understanding of chemical toxicity was incomplete, these observations signaled that petroleum's presence was more than a nuisance—it might pose genuine health risks.

A few physicians advocated for pumping standing oil from neglected pits to reduce mosquito breeding, or for relocating refineries away from residential clusters. Their proposals, though practical, demanded cooperation from local authorities or well owners who did not always see the urgency. Nevertheless, these medical voices formed part of the patchwork of early environmental concern, bridging anecdotal patient complaints with broader public health considerations.

7. Conserving Resources or Maximizing Profit? Competing Mindsets

7.1. Immediate Extraction Versus Long-Term Stewardship

Many 19th-century oil operators prioritized short-term gains, mindful that wells could run dry or that new technology might make small operations obsolete. They had little incentive to invest in long-term resource stewardship, especially when faced with tight capital. If burying waste cheaply or draining seepages quickly turned a faster profit, that often took precedence over the intangible benefits of environmental care.

However, a handful of landowners recognized that reckless extraction might spoil farmland or degrade water sources crucial for future generations. They argued for gentler methods—limiting well depths, plugging dried holes, or rotating drilling to avoid overtaxing local geology. Such voices tended to be wealthier, with the luxury to think about the land's long-term viability. Their stance foreshadowed modern sustainability arguments, albeit practiced on a small scale and lacking broad support.

7.2. Conflicts Between Small Landowners and External Investors

As the century advanced, outside investors began taking more interest in fledgling oil fields. They brought capital for deeper wells or bigger refineries, but their remote ownership often cared little about local ecological health. Small landowners who lived on-site bore the brunt of pollution or water shortages if external operators cut corners. Tensions flared in community meetings, with local families accusing absentee investors of ignoring the region's well-being.

Sometimes, these disputes ended with outside financiers quietly paying minimal settlements or bribes to local officials. Other times, local activism prevented new wells from being drilled unless the operator promised certain safeguards. In essence, these confrontations underscored a core clash: the push for rapid industrial exploitation of oil versus the desire of on-the-ground residents to preserve their environment and health. This dynamic mirrored other resource frontiers of the 19th century, like timber or minerals, but with petroleum's distinct set of messy byproducts.

8. Animal and Wildlife Impacts

8.1. Oil Pits as Hazards to Livestock

Farm animals occasionally wandered into unprotected oil pits. Horses or cows became stuck in thick sludge, sometimes suffocating or injuring themselves. Farmers who lost valuable livestock demanded compensation, inciting local disputes or even violent reprisals if the pit owner refused. Fear of further accidents led some communities to require fences or watchmen around large pits.

For smaller operators, these incidents were financially damaging, as they had to pay for the lost animal or risk communal wrath. Repeat offenders who neglected basic precautions might be run out of town or forced to shut down. In this sense, livestock accidents served as an early impetus for minimal safety measures, even in places where environmental consciousness was otherwise low.

8.2. Effects on Fish and Aquatic Life

Though systematic ecological studies were rare, anecdotal evidence suggested that oily films on rivers harmed fish populations. Local fishers reported reduced catches downstream of refineries or well clusters. Some observed that fish developed lesions or died off in large numbers after heavy rainfall washed fresh oil residues into the water.

Concerns over losing fish—a critical food source or livelihood—led fishermen to protest directly. They might blockade transport routes or sabotage equipment until authorities intervened. In response, a handful of local governments in Europe and North America instituted seasonal restrictions on certain oil-dumping practices, hoping to protect spawning periods. While not framed as modern "conservation," these measures aimed to preserve wildlife that communities relied upon, again reflecting the era's practical rather than philosophical environmental orientation.

9. Notable Environmental Incidents and Public Outcry

9.1. The "Galician Spill" of the 1840s

Documents from the mid-1840s reference a significant spill in one of Galicia's emerging oil-producing districts. Heavy rains collapsed an earthen reservoir used to store oily waste, sending a large flow into a nearby river that watered multiple villages. Fishermen found entire stretches of river coated in a rainbow sheen, and the smell lingered for weeks. Many local families blamed dead fish and farmland contamination on the spillage.

Newspapers from nearby towns covered the incident, amplifying calls for stronger oversight. The local Austrian administrator conducted an inquiry, eventually fining the reservoir's owner and ordering him to build a sturdier containment structure. Though no broad legislative changes followed, the event served as a cautionary tale. Merchants considering oil expansion in the region realized that major mishaps could lead to forced shutdowns, community anger, and personal liability.

9.2. North American Well Blowout Cases

In North America, a few well blowouts in the 1850s spewed oil over surrounding fields, saturating soil and polluting small creeks. While these blowouts were modest compared to later decades, they still shocked local residents unused to seeing thousands of gallons of oil spurt uncontrollably. Some blowouts caught fire, creating towering flames and black smoke visible for miles, spurring rumors of a "volcanic eruption."

Public outcry followed: farmers demanded compensation for ruined crops, while local preachers used the calamity in sermons, warning about the dangers of human greed tampering with the earth. These episodes shaped community norms, prompting well owners to adopt slightly safer drilling methods, line their boreholes, or keep better firefighting supplies on hand. Though hardly foolproof, such incremental changes reduced the frequency of catastrophic blowouts, illustrating how direct environmental crises forced improvements in industry practices.

10. Seeds of Broader Environmental Ethics

10.1. Philosophical Underpinnings

While the idea of protecting nature purely for its own sake was rare in the early 19th century, certain philosophers and writers did champion more harmonious human-nature relations. In Europe, Romantic poets exalted untouched landscapes; in the United States, figures like Henry David Thoreau emphasized the spiritual value of wilderness. Although these thinkers seldom wrote specifically about oil, their broader calls to respect the natural world resonated with some communities dealing with pollution.

A few forward-looking individuals drew connections between the moral or aesthetic reverence for nature and the practical harm inflicted by careless industrial activity. They argued that humankind risked losing beauty, health, and moral clarity if it systematically destroyed local ecosystems for profit. Among these early voices, some singled out the oil sector as an example of how short-sighted exploitation could degrade entire regions.

10.2. First Efforts at Organized Complaint Groups

In a handful of places, small groups formed—akin to proto-environmental committees—focused on issues like water cleanliness or farmland preservation. Sometimes they rallied around multiple concerns: sawmill runoff, sewage from expanding towns, and oil-related pollution. Their membership might include farmers, local gentry, or even sympathetic business owners who recognized that a well-kept environment was key to long-term prosperity.

Though these groups had limited reach and power, they occasionally persuaded local boards to pass stricter ordinances or to enforce existing rules more vigorously. Their activities foreshadowed the civic activism that would become more prominent in later decades, once industrialization's harms grew more visible. For 19th-century oil communities, these committees offered one of the few structured channels to address pollution or reckless extraction practices collectively.

11. Lasting Legacy and Transition to Late-Century Developments

11.1. Evolving Public Consciousness

By the latter half of the 19th century, people in regions with any significant oil activity were increasingly aware that mishandling the resource could yield real damage. The concept of "environmental protection" was not fully formed, but local survival instincts, health concerns, and economic interests converged to create a patchwork of informal and minor formal rules. Where communities forced accountability, operators made small concessions to safety or waste management. Where local leadership was lax, contamination continued relatively unchecked.

Over time, these experiences fed into a budding sense that industrial activities—including oil—demanded oversight. As technology advanced, larger wells and more sophisticated refineries accentuated the stakes. If a single operation could produce more oil in a month than earlier pits had in a year, the potential for environmental harm multiplied. Thus, even as some mid-century concerns were patched up, new frontiers of extraction threatened larger-scale impacts, setting the stage for deeper environmental debates by century's end.

11.2. Prefiguring Modern Environmental Regulation

The modest regulations, local ordinances, and community-led initiatives formed the early building blocks of what would later become recognized as environmental law and activism. Although far from comprehensive, these scattered efforts showed that people would not always tolerate unconstrained pollution for the sake of profit. They demonstrated that local economies and ecosystems could be deeply intertwined—and that ignoring one could undermine the other.

When major oil booms arrived in the late 19th century (and especially into the 20th, beyond our scope here), the lessons gleaned from these earlier environmental mishaps and criticisms would resurface. Over decades, they informed more robust legislation, scientific monitoring, and advocacy movements. In that sense, the 19th-century experiences, though limited, were crucial in shaping the initial ethos around balancing oil extraction with communal and ecological well-being.

CHAPTER 19: GLOBAL SEARCHES FOR OIL IN THE LATE 19TH CENTURY

Introduction

By the last few decades of the 19th century, oil had emerged from scattered local use to become a commodity sought by an increasingly interconnected world. Earlier chapters showed how communities discovered oil seepages, tested drilling methods, and found ways to refine and market the products. Yet, much of this activity remained confined to specific regions—North America, parts of Eastern Europe, the Caucasus around Baku, and a few other locales.

During the late 1800s, the drive to find new oil fields intensified. Armed with incremental drilling improvements and a rising awareness that petroleum could serve as both a lamp fuel and a mechanical lubricant, explorers, entrepreneurs, and speculators began scouring distant lands, hoping to discover "the next big field." Some expeditions relied on rumored seepages or scattered geological reports, while others combined scientific surveys with good old-fashioned luck. This chapter examines these global searches for oil, their successes and misfires, and how they laid the groundwork for a truly international oil industry as the century drew to a close.

1. The Shifting Context of the 1870s and 1880s

1.1. Growing Demand for Lighting and Lubricants

By the 1870s, the appetite for refined lamp oil had grown considerably in urban centers. Whale populations had been diminished by extensive hunting, making whale oil scarcer and more expensive. Meanwhile, industrializing nations needed better lubricants for machinery, especially as railways spread and factories grew. These dual forces—lighting and lubrication—drove up petroleum's perceived value. Despite improvements in coal gas for city street lighting, many households and small businesses still relied on oil lamps, especially outside major metropolitan areas.

Additionally, shipping companies and emerging industrial complexes recognized that petroleum-based lubricants outperformed older greases. While the volumes involved were not as vast as they would become in later decades, the steady rise in usage was enough to encourage prospectors, financiers, and governments to wonder if more large oilfields lay undiscovered in distant lands.

1.2. Technological Advancements in Drilling

In previous chapters, we discussed how cable-tool methods, adapted from salt drilling, proved slow and laborious. The late 19th century saw gradual refinements: improved cable materials, sharper drill bits, and somewhat better well casings to prevent collapses. While these innovations were not revolutionary, they allowed operators to sink wells deeper and faster, broadening the search zones beyond shallow seepages.

Simultaneously, geologists refined their understanding of sedimentary layers, correlating certain rock types—like porous sandstones or limestones—with potential oil accumulation. Though far from foolproof, these insights helped direct drilling efforts toward more promising strata rather than random guesses. As a result, explorers felt emboldened to attempt wells in regions previously dismissed as unworkable.

1.3. International Travel and Communication

Global travel and communication networks became more extensive in the late 1800s. Steamships traversed oceans on predictable schedules, telegraph lines connected continents, and railway expansions in Asia, Africa, and Latin America opened up once-remote frontiers. This infrastructure meant that news of a successful gusher in the Caucasus or a promising seep in Sumatra could reach London, Paris, or New York within days. Potential investors could mobilize quickly, financing expeditions or forming syndicates to exploit rumored deposits half a world away.

As competition grew, national governments also took notice, realizing that local oil discoveries might enrich the state treasury or enhance domestic industry. Together, these technological and economic shifts sparked a flurry of exploratory ventures across the globe, each hoping to uncover significant oil reserves before rivals did.

2. North America's Westward Oil Explorations

2.1. Beyond Pennsylvania: Kentucky, Ohio, and Other States

Pennsylvania's Drake Well (struck in 1859) is often cited as the beginning of the modern U.S. oil industry, but by the 1870s, productive wells had appeared in Kentucky, Ohio, West Virginia, and even parts of Illinois and Indiana. Some states, initially overshadowed by Pennsylvania's fame, quietly developed mid-scale fields, fostering local refineries and modest pipelines.

Entrepreneurs roamed the American interior, buying mineral rights from farmers who had no idea whether black gold lay beneath their cornfields. Small towns boomed if a well hit, filling with hurried speculators, rig builders, and merchants eager to profit from every aspect of an oil rush—from lodging to drilling supplies. Yet many of these fields delivered less output than hoped, and the typical pattern of "boom and bust" repeated. Once the wells declined, the horde of speculators moved on, leaving behind ghost-town relics of the fervor.

2.2. The Emergence of Pipeline Networks

One notable development in North America during this period was the gradual spread of pipeline infrastructure, especially in states like Pennsylvania and Ohio. Rather than relying solely on barrel-hauling wagons, operators began laying pipelines to carry crude directly from wellheads to nearby rail stations or refineries. Although early pipelines were prone to leaks and sabotage by rivals, they demonstrated a more efficient, cost-effective way to transport oil.

This innovation not only reshaped local logistics but also set a precedent for searching fields further away from major roads or waterways. Previously, a remote site might be deemed impractical if barrels had to travel dozens of miles over rough terrain. Pipelines partly overcame these obstacles, encouraging deeper ventures into frontier lands. Those with capital to invest in pipeline construction could tap previously unreachable reserves, reinforcing the sense that North America's oil bounty extended well beyond Pennsylvania's earliest discoveries.

2.3. Influence on Settlement Patterns

As new wells appeared, local economies near them diversified. Merchants arrived to supply rigs, blacksmiths specialized in forging drilling tools, and carpenters built derricks. Entire families relocated, gambling that a fresh discovery would yield stable incomes. In states like Texas—though still in the nascent stage of oil exploration during the late 19th century—some geologists and prospectors surveyed promising geological formations, anticipating big finds that would eventually emerge in the next century.

While the large-scale Texan booms remained just over the horizon, the seeds were planted. Early references in local newspapers around 1880 mention seeps or "faint indications" of deeper oil-bearing strata, fueling speculation that the southwestern U.S. might eventually rival Pennsylvania's success. Thus, the late 19th century in North America was marked by a dynamic interplay of proven fields, modest expansions, and rumors pointing to even grander potential in the country's vast interior.

3. Eastern Europe and Russia: The Rise of Baku and Beyond

3.1. Baku's Rapid Growth in the Late 1800s

Earlier chapters covered Baku's historical usage of shallow pits and modest wells. By the 1870s and 1880s, however, the region's oil industry soared. Entrepreneurs—both local and foreign—recognized the potential for deeper drilling, aided by improved rigs. Combined with the Russian government's willingness to grant concessions, capital flowed into Baku. This influx of money and machinery transformed the area into a bustling hub of derricks, refineries, and infrastructure projects.

One significant advantage was Baku's location near the Caspian Sea. Barges or small tankers could ship crude across the water, linking Baku to broader markets. Local banks financed expansions of piping networks to carry oil from well clusters to refineries. By the mid-1880s, some refineries in the Baku area rivaled those in North America for scale. Although output still lagged behind the largest American fields, the rapid rise turned heads across Europe, reinforcing the idea that major oil basins existed outside the New World.

3.2. Galicia's Continued Development

In Austrian Galicia, modest expansions continued through the late 1800s. Towns like Drohobycz and Borysław became local centers for drilling and refining. While overshadowed by Baku's scale, Galician fields supported a steady production that supplied Central Europe with lamp oil and lubricants. Eastern rail networks facilitated transport, linking the region to cities like Vienna or Budapest.

Multiple small operators thrived here, though conflicts arose over land rights and resource disputes. Some synergy emerged between local drilling and the area's established refining craft, fostering incremental technological improvements. By the century's end, Galicia had solidified its reputation as a reliable, if not spectacular, European oil province—one that would see further expansions in the early 20th century (though that lies beyond our current scope).

3.3. Roving Speculators in the Caucasus and Eastern Territories

Baku's success enticed speculators to roam other parts of the Caucasus and southwestern Russia, scouring for seepages or geological indicators. Some geologists posited that the same sedimentary formations fueling Baku's bounty might extend into neighboring areas. Exploratory wells sprouted along the Caspian coastline and inland valleys. Although many turned up dry or yielded minimal flows, a few moderate discoveries reaffirmed optimism about the region's potential.

Russian authorities, for their part, balanced granting concessions to foreign investors with maintaining national oversight. While no sweeping laws tightly regulated the new fields, local governors recognized that large foreign-run operations could overshadow local efforts. Hence, they demanded partial partnerships or fees that funneled revenue back into imperial coffers. By 1900, the Caucasus had become a magnet for adventurous drillers, helping shape Russia's broader industrial ambitions.

4. The Middle East: Early Forays and Limited Results

4.1. Ottoman and Persian Territories

Stories of "burning springs" and natural asphalt deposits in regions like Mesopotamia or southwestern Persia circulated among European travelers for centuries. In the late 19th century, spurred by Baku's dramatic rise, scattered Western expeditions ventured into these lands, hoping to replicate the Caucasus success. Some obtained concessions from local Ottoman or Qajar Persian officials, while others negotiated deals with tribal leaders who claimed authority over remote areas.

However, these early attempts often struggled with logistical and political hurdles. Transportation networks were sparse, local governance was complex, and the geology—though promising—was less understood than that of the Caucasus. A few shallow wells drilled in southwestern Persia encountered small pockets of oil, but production remained negligible on a world scale. Observers recognized that large discoveries might be possible but concluded that deeper drilling and more stable political conditions were required. Thus, while the seeds of future Middle Eastern oil booms were planted, the late 19th century saw only preliminary steps in that direction.

4.2. British Interests and Survey Missions

The British Empire, concerned primarily with strategic routes to India and maintaining trade dominance, dispatched occasional survey teams to evaluate resource potential along the Persian Gulf. Some explorers, half-diplomat and half-geologist, wrote back to London describing bitumen-laced coasts or seepages in mountainous interior zones. They cautioned that developing these resources would demand significant infrastructure and more advanced drilling rigs.

Because Britain's immediate colonial focus lay in controlling maritime choke points, systematic oil exploitation took a back seat. Nonetheless, the intelligence gathered in these missions formed the basis of future negotiations. By century's end, a few British entrepreneurs quietly held "paper concessions" for tracts of land, anticipating a time when conditions ripened for serious drilling. This approach mirrored the empire's broader pattern: secure strategic footholds first, ramp up resource extraction once political or technical obstacles subsided.

4.3. Local Uses and Perceptions

In many Middle Eastern villages, residents continued using surface bitumen for waterproofing or trade, as they had for generations. Some communities, hearing about foreign curiosity in the "black substance," reacted with skepticism or worry, especially if it heralded increased outside interference. Others hoped foreign capital might improve local roads or create jobs. In general, though, the late 19th century saw no dramatic oil rush here, merely glimmers of future potential overshadowed by more immediate imperial and regional concerns.

5. Southeast Asia and the Dutch East Indies

5.1. Indonesian Archipelago's Natural Seepages

The Indonesian islands (then part of the Dutch East Indies) had long been known for their spice wealth. Less documented, but still present, were scattered oil seeps or tar springs. By the 1870s, some Dutch colonial officials and private explorers began paying attention. If new drilling technology or refining methods could be brought here, they speculated, the archipelago might supply an extra revenue stream for Dutch colonial coffers.

However, the logistical challenges were immense. Thick jungles, limited road networks, and local resistance to forced labor or land seizure all impeded large-scale drilling. Thus, only a few modest wells appeared, often near known seeps, producing enough crude for local lamp oil or minor exports. The Dutch authorities mostly considered these efforts an experiment, overshadowed by the region's more established plantation economies (coffee, sugar, rubber). Yet, like the Middle East, this region's geology hinted at deeper reservoirs, waiting for more determined exploitation.

5.2. Sumatra and Early Joint Ventures

Sumatra, in particular, drew a handful of foreign entrepreneurs who formed small joint ventures with local Dutch officials. They raised capital in Amsterdam or The Hague, imported secondhand cable-tool rigs, and hired local laborers. Drilling success was mixed—some struck modest flows, enough to justify small refining setups near the coast. But these projects rarely soared to the scale of Baku or Pennsylvania.

Nonetheless, the presence of even a handful of operating wells signaled that Southeast Asia could be added to the growing map of potential oil provinces. Merchants shipping Sumatran oil to Singapore or other regional ports occasionally reported fair profits. The entire enterprise was embryonic, but it stood as proof that the global search for oil included corners of Asia often associated primarily with spices and agricultural exports.

6. The Role of Geologists and "Oil Scouts" in Late 19th-Century Exploration

6.1. Scientific Surveys and University Influence

By the 1880s, geology had matured as a discipline, with universities in Europe and North America teaching structured courses on stratigraphy, fossil analysis, and mineral exploration. Some newly trained geologists specialized in "petroleum geology," applying field mapping and rock correlation to identify likely oil-bearing horizons. This scientific approach improved success rates somewhat, though wildcat drilling (drilling in unproven areas on a hunch) still dominated.

When an oil-seeking expedition left for Asia, Africa, or Latin America, it might include a geologist who surveyed surface outcrops, collected samples, and recommended where to sink test wells. The repeated reliance on such specialists reflected a growing confidence that geology was not just academic theory, but a practical tool for resource discovery. Their presence elevated exploration beyond mere guesswork, fueling the era's sense of methodical resource expansion.

6.2. Adventurous "Oil Scouts" and Rumor-Based Quests

Alongside the formally trained geologists, a motley group of self-styled "oil scouts" roamed the planet. These individuals, sometimes lacking formal education, relied on experience in prior booms and a keen ear for local rumors. They traveled to remote districts with minimal equipment, bargaining for land rights if they smelled a hint of oil seepage or if local folklore mentioned "eternal fires."

Their adventures, often recounted in letters or sensational newspaper interviews, combined tall tales with genuine prospecting. Some scouts found modest fields, flipping the land to larger companies for profit. Others wasted months chasing myths. Nonetheless, the existence of these scouts underscored that the chase for new oil was no longer confined to known districts—every corner of the globe seemed fair game if the conditions hinted at subterranean riches.

7. Financial Syndicates and International Investment

7.1. Emergence of Oil Companies and Joint-Stock Ventures

During the late 19th century, more organized oil companies replaced smaller family-run or individual operations, especially in the United States and parts of Europe. Investors pooled capital in joint-stock ventures, distributing risk among multiple shareholders. This model let them finance larger drilling campaigns in remote regions, pay for geological surveys, and even build rudimentary infrastructure once a well showed promise.

A prime example: In the U.S., integrated companies that combined drilling, refining, and distribution realized economies of scale. While the famous rise of certain oil trusts belongs partly to the next century (beyond our scope), the seeds were sown here. Foreign investors also took note, buying shares in American or European companies with active drilling projects. Such cross-border financing fueled a more international perspective, with shareholders in London or Paris profiting from a well in Ohio or Baku, or from an exploratory mission in Sumatra.

7.2. Speculative Bubbles and Hype

Hand in hand with genuine investment came speculative mania. Newspapers occasionally hyped rumored "monster strikes," driving up share prices of small exploration companies—only for the rumor to fade when the well yielded minimal oil. Some unscrupulous promoters salted land with bits of black tar or manipulated geological data to entice wealthy but uninformed backers. The resulting booms and busts mirrored patterns seen in earlier resource rushes (e.g., gold or silver).

Governments in a few countries began passing light regulations on stock offerings, requiring minimal disclosures about drilling sites, though enforcement was patchy. Potential investors had to weigh the risk of losing everything against the chance of striking an oil fortune. These financial swells reflected the broader climate of late 19th-century capitalism: dynamic, globalizing, and sometimes fraught with speculation. By extension, the oil sector showcased both the promise and the peril of resource investment in a newly connected world.

8. Challenges of Transport and Refining in Far-Flung Districts

8.1. Building Infrastructure from Scratch

Exploring remote regions for oil meant more than just drilling. Without roads, rail lines, or ports, even a productive well was worthless if operators could not ship the crude. Some companies, flush with capital, built rudimentary roads or small gauge rail spurs to haul barrels to a navigable river or seaport. Others established local refineries to convert crude into lamp oil or heavy lubricants on-site, reducing the volume to transport.

These undertakings were risky. An area might yield oil initially, prompting expensive infrastructure, only for the main reservoir to run dry or produce far less than anticipated. Project backers sometimes lost fortunes in half-built pipelines or idle refineries. Conversely, if an operation succeeded, it might anchor a new regional economy, forming the skeleton for future expansions. This dynamic—high risk, high reward—characterized the global push to find oil beyond well-known basins.

8.2. Refining Quality Variations

Transporting unrefined crude over long distances was costly, so local refining often made sense. However, the quality of local refining varied widely in remote areas. Skilled labor was scarce, and the technology arrived in crates, subject to mishandling or incomplete assembly. Equipment might break down in humid tropical climates or rust in desert conditions. The resulting oil products could be contaminated or poorly fractionated, limiting their market appeal.

Companies that overcame these hurdles and produced consistent lamp oil or lubricants found appreciative buyers in nearby towns or cities lacking domestic supplies. Even so, the learning curve was steep, and repeated failures were common. The synergy between drilling, refining, and distribution demanded robust organization—foreshadowing the integrated approach that would dominate in the 20th century.

9. Governmental Policies and Colonial Ambitions

9.1. Concessions and Diplomatic Maneuvering

In colonial regions—such as parts of Africa, Southeast Asia, or the Middle East—imperial governments typically held sweeping powers over resource extraction. They might grant concessions to private companies, extracting royalties or taxes in return. These deals were often negotiated via diplomats or colonial administrators who had limited geological knowledge but recognized the potential for boosting local revenue or strategic footholds.

In some cases, local elites or sultans cooperated with foreign explorers, hoping to modernize their territories or glean personal benefits. Other times, they resisted, fearing exploitation and environmental harm. Diplomatic tension arose if multiple foreign powers vied for the same region's rumored oil fields, each pressuring local rulers for exclusive rights. Though full-blown oil wars remained a future phenomenon, these smaller conflicts hinted at how oil would eventually intertwine with global politics.

9.2. Protective Tariffs and National Supply

For certain nations, controlling or subsidizing domestic oil sources was deemed vital for self-sufficiency, especially for fueling lighthouses, rail networks, and the navy's needs. Hence, governments occasionally imposed tariffs on imported petroleum products to protect local refiners. Alternatively, they might offer tax breaks or incentives to spark drilling campaigns in their territory, hoping to reduce reliance on foreign oil.

These policies had mixed effects. Some spurred local industries that otherwise struggled to compete with cheaper foreign oil. Others led to price distortions or invited corruption, as unscrupulous operators gamed the system. Still, the broader implication was clear: by the late 19th century, states recognized that oil had strategic and economic weight, justifying policy interventions to secure reliable supplies.

10. Cultural Interactions and Local Communities

10.1. Indigenous and Rural Responses

In remote corners of Asia, Africa, or Latin America, foreign-led expeditions seeking oil encountered local societies with their own land-use traditions. Drilling or building a refinery might disrupt long-standing agricultural or pastoral systems, fueling resentment. Some communities, fearful of losing farmland or polluting water, resisted outside surveyors, occasionally resorting to sabotage or open hostility.

On the flip side, a few local leaders welcomed the money and modern goods that foreign companies brought. They negotiated terms that granted them a share of profits or guaranteed improvements like wells for drinking water or small clinics. The success of these arrangements varied widely: in certain places, genuine partnerships formed, while in others, local people felt cheated by lopsided deals that favored foreign investors. This dynamic mirrored the broader colonial pattern of resource exploitation, with oil as one more factor in the complex interplay of cultural contact and economic transformation.

10.2. Exchange of Knowledge and Techniques

When drilling teams established camps in far-flung regions, they inevitably trained local workers to operate cable-tool rigs, build derricks, or carry out refining steps. Over time, this exchange of skills allowed some indigenous laborers to become proficient in the technical aspects of petroleum extraction. A few rose to supervisory roles, bridging cultural gaps between foreign managers and local labor forces.

Additionally, local farmers or artisans who learned about petroleum's properties might incorporate it into existing customs: using it to seal wooden items, repel insects, or craft small lamps. Though overshadowed by the export focus of these foreign companies, these small cultural adaptations underscored how, even in the late 19th century, oil was not purely an external imposition but a substance that could merge with local practices.

11. The Stage Set for the 20th Century

11.1. Conclusions from Late 19th-Century Searches

By the final years of the 1800s, the global map of known oil provinces had expanded beyond the pioneer fields. North America boasted multiple states with proven reserves, the Russian Empire's Caucasus region became a major producer, Eastern Europe's Galicia held steady output, and scattered fields in Asia hinted at further potential. Nations and investors had begun forging policies, concessions, and infrastructure to harness these finds.

Despite many dry holes and financial disasters, the overall trend was positive—petroleum production was rising, feeding industrial modernization. The world's appetite for refined products likewise grew, as new lighting demands and complex machinery required reliable lubricants. The synergy between drilling, refining, and transport networks advanced steadily, albeit unevenly, across continents.

11.2. Paving the Way for Accelerated Growth

With each successful discovery came improved technology, bigger ambitions, and fresh impetus for exploring new territories. Entrepreneurs in one region learned from the mistakes in another, geologists refined their theories, and pipeline or shipping methods advanced. As the 19th century ended, the foundation was laid for the explosive oil developments of the early 20th century—an era we do not cover in detail here but one that undeniably benefited from these late 19th-century forays.

Thus, the global searches for oil in the last decades of the 1800s not only shaped individual fields and regional economies; they united the entire concept of "oil exploration" under an international framework. Drilling rigs, capital, and scientific expertise flowed across borders in ways rarely seen before, turning a once-localized resource into a budding international commodity. In the next (and final) chapter, we will examine how this transitional moment—right at the close of the 19th century—placed humanity on the verge of a new oil-driven epoch, without stepping into the full modern era. The threshold had been reached; the next step would be to cross it.

CHAPTER 20: THE THRESHOLD OF THE MODERN OIL ERA

Introduction

By the end of the 19th century, the slow but steady evolution of the oil industry had reached a pivotal juncture. From humble beginnings—where people scooped bitumen from seeps or tested shallow wells in backwater towns—petroleum had become a recognized commodity, supporting growing markets in lighting, lubrication, and small-scale industrial uses. Several regions produced oil on a noteworthy scale, a few large companies were forming, and a global sense of possibility now enveloped the resource.

This final chapter synthesizes the major trends we have traced throughout the book. We will look at how late 19th-century improvements in drilling, refining, and transport converged to push oil beyond a minor resource and into a position of far-reaching influence. Although we will not delve into the full scope of the 20th century, we will outline how the stage was set for oil to become an unparalleled economic and geopolitical force in modern times. In doing so, we remain within the historical context up to the cusp of the 1900s, showing how each piece—technology, policy, social acceptance, and environmental awareness—aligned to create an industry on the brink of dramatic change.

1. Recap of Oil's 19th-Century Journey

1.1. From Local Curiosity to Recognized Commodity

In the early 1800s, oil was largely a regional affair—used in rural communities or overshadowed by whale oil and coal for industrial power. Bit by bit, improvements in drilling (cable-tool rigs, deeper wells), refining (distillation techniques), and distribution (basic pipelines, barrel transport) transformed it into a viable commercial product. By the 1870s and 1880s, major fields like Pennsylvania in the U.S. and Baku in the Russian Empire demonstrated petroleum's ability to supply lamp oil and lubricants at scale, spurring further exploration around the world.

Public perception shifted along the way. What had once been a smelly, marginal substance found in remote seepages evolved into a resource that newspapers discussed frequently, that local communities built small economies around, and that scientists recognized as geologically significant. The combined effect was to lift oil out of obscurity, giving it real weight in industrial and societal development.

1.2. Diverse Applications Driving Demand

Though coal still dominated large-scale power generation, oil found strong footholds where it excelled: lamp fuel for domestic and street lighting outside major city gas grids, heavy lubricants for factory machines, and specialized uses in certain maritime contexts. Patent medicines and folk remedies also contributed to demand, albeit on a smaller scale. The synergy of these varied applications guaranteed a stable or even rising market price—encouraging drilling expansions in multiple locales.

Refiners who produced better-quality lamp oils gained local or regional reputations, hooking customers with a cleaner burn and less odor. Industrial buyers, especially railway and steamship operators, valued the consistency and high-temperature performance of petroleum lubricants. These steady markets underpinned oil's economic viability, giving entrepreneurs confidence to take bigger exploration risks.

2. The Maturing Drilling and Refining Industries

2.1. Standardized Equipment and Methods

By the late 19th century, the spread of standardized drilling components reduced the chaos of earlier trial-and-error approaches. Manufacturers produced cable-tool rigs, drill bits, and pumps according to recognized specifications, making it easier to ship equipment worldwide. Drillers from Pennsylvania, for instance, could relocate to Eastern Europe or the Caucasus and find rigs that worked in familiar ways, requiring minimal adaptation.

Refiners likewise benefitted from more uniform apparatuses. Steel or iron stills replaced crude copper kettles in many operations, improving safety and allowing better temperature control. Certain chemical wash steps—such as treating oil with acid or alkali—became standard, yielding clearer fractions. This consistency reassured customers that a given brand of lamp oil or lubricant would behave predictably, spurring brand loyalty and broader acceptance.

2.2. Growth of Integrated Operations

While many small operators still existed, some wealthier individuals recognized the advantage of integrating drilling, transport, refining, and distribution under one organizational umbrella. This strategy, tested in limited form earlier, gained traction in the late 1800s. By controlling each stage, companies could reduce costs, ensure consistent quality, and respond swiftly to market shifts.

Though the full "vertical integration" typical of 20th-century oil giants was still forming, the seeds were present. In North America, a few well-financed groups began consolidating smaller producers, building or buying pipelines, and acquiring refineries. Similarly, in places like Baku, certain entrepreneurs and bankers combined their interests to create stable supply chains from wellhead to shipping port. These organizational innovations laid the groundwork for more cohesive, large-scale enterprises that would dominate the next century's petroleum landscape.

3. Social and Cultural Shifts at the Century's End

3.1. Oil as a Symbol of Progress

In industrializing societies, petroleum came to symbolize modernity and efficiency. Newspapers lauded new lamps that replaced dimmer or smellier alternatives. Factories boasted of improved output because of machine lubricants that reduced breakdowns. Towns that struck oil sometimes rebranded themselves as forward-looking, erecting better roads or public buildings financed by oil taxes or philanthropic well owners.

Urban households with refined petroleum lamps perceived themselves as keeping pace with the times, akin to adopting the latest stove or sewing machine. This mindset fed an undercurrent of optimism about oil's potential to remake society for the better—improving health, boosting production, and brightening homes. Yet beneath that optimism lurked reminders of accidents, pollution, and the hazards of speculation, ensuring that oil's image remained complex.

3.2. Evolving Worker Experiences in Oil Fields

As some fields expanded, labor conditions became more structured. A workforce of drillers, rig assemblers, pipeline fitters, and refinery hands formed semi-stable communities around clusters of wells. Small towns sometimes developed rudimentary amenities—boarding houses, saloons, a school—catering to these transient laborers. Over time, a portion of workers settled down, raising families in emerging "oil patches" across the U.S. interior, the Caucasus, or Eastern Europe.

Workers' social lives centered on these new communities. They swapped stories of near misses with blowouts or recounted pay disputes. Occasionally, they organized to demand safer conditions, but such efforts remained scattered due to the rotating nature of the workforce. Still, a shared identity slowly emerged: toiling in the oil sector demanded grit, mechanical know-how, and acceptance of risk. This identity fostered camaraderie and set oil workers apart from other industrial laborers.

4. Political and Environmental Underpinnings

4.1. Early Geopolitical Maneuvers

While the late 19th-century push for resources mostly centered on coal, iron, and farmland, governments did not ignore oil, especially if it factored into national self-reliance or potential naval uses. States with oil-producing regions—like the Russian Empire—noticed the revenue potential. Colonial powers eyed new territories with rumored seepages or promising geology, occasionally jockeying for concessions.

No major conflicts erupted solely over oil in this period, but small-scale diplomatic tussles and local tensions set a precedent for resource-driven politics. Land treaties in places like Persia or the Ottoman Empire included references to bitumen deposits, with foreign advisors quietly urging sultans or shahs to sign favorable deals. These subdued actions foreshadowed the full-blown geopolitical maneuvers that would define the next century, but they also showed that oil was already a strategic consideration among empire-builders.

4.2. Intensifying Environmental Worries

Chapter Eighteen highlighted local protests and basic regulations to curb pollution. As more fields opened, these issues multiplied. Oil spills during transport, open-air refining, and large-scale dumping took a toll on streams, farmland, and air quality. A few philanthropic well owners tried replanting areas damaged by drilling or paying for cleaner disposal methods. Such gestures were sporadic, overshadowed by the lure of profit.

Observers increasingly realized that the old methods of unlined pits, casual burning, and minimal safety oversight might not sustain the resource or local ecosystems for the long haul. While a structured environmental movement had not yet crystallized, newspapers and local committees continued pressing for stricter controls in major production zones. In the slow, piecemeal manner typical of 19th-century governance, these demands occasionally yielded modest protective measures—early glimpses of how the modern era would grapple with balancing oil output and ecological health.

5. Transition in the Broader Energy Landscape

5.1. Continued Role of Coal and Electricity's Emergence

Despite oil's advances, coal still powered the vast majority of steam engines, from locomotives to factory boilers. Meanwhile, the late 19th century saw the rise of electrical systems for city lighting, overshadowing some roles that oil-based lamps had filled. This forced refiners to adapt: while they lost a slice of the lighting market to electric grids in major urban centers, they found stable demand in smaller towns and the countryside, as well as in lubrication.

Moreover, industrial machinery still depended heavily on lubricants, an area where electricity had little immediate competition. Thus, oil did not replace all earlier fuels or transform the entire energy sector at once. Instead, it carved out a niche, steadily expanding into multiple corners of the economy, from household lamps to specialized mechanical applications, ensuring that its foothold would remain firm despite evolving energy competition.

5.2. Early Whisperings of Internal Combustion

By the 1890s, inventors in Europe tinkered with internal combustion engines that burned gasoline or similar light oil fractions. While such engines were few and far between—and not yet overshadowing horse-drawn transport—reports of experimental automobiles or stationary engines reached newspapers, fueling speculation about a new frontier for petroleum. The volumes required for these prototypes were minuscule, so they did not affect mainstream drilling or refining.

However, the concept that oil-based fuels might power small engines with greater efficiency than steam lingered on the fringes of industrial discourse. A few visionaries quietly predicted that, if the technology matured, it could revolutionize mobility. Though outside the strict purview of our 19th-century focus, these ideas illustrated a bridging moment: oil was no longer confined to lamps and mechanical lubrication. In a not-too-distant future, it might energize an entirely new sector of transportation and industry, ushering in transformations still unimaginable to most people in the late 1800s.

6. Late-Century Market Structures and Emerging Titans

6.1. Regional Dominance and Price Fluctuations

Markets for oil remained unstable, prone to local gluts or shortfalls. If new wells in Pennsylvania or Baku spiked production, prices sometimes plummeted, ruining smaller producers. Conversely, major accidents or political unrest could curtail supply, causing sharp price climbs. This volatility led some operators to seek ways to stabilize or coordinate production, planting the seeds of quasi-cartel behaviors.

North America saw the rise of powerful refining and pipeline networks that leveraged their scale to negotiate favorable terms with well owners and railroad companies. Similar patterns occurred in Baku, where a few large outfits started to overshadow smaller drilling groups. Although the next century would see more formal oil trusts and cartels, the kernel of centralized control emerged in the 1890s, as big players learned to wield their logistical and financial strength.

6.2. Shaping Public Policy

Local and national governments also began gleaning lessons about how to regulate or tax the sector. A stable oil supply conferred economic benefits, so some states offered mild incentives to refine or transport petroleum. Others, worried about foreign control, limited how much equity foreign investors could hold in strategic oilfields. These early policy maneuvers aimed to guard national interests without stifling growth.

Public acceptance of such interventions was mixed. Certain businesspeople resented state meddling, preferring a laissez-faire approach. Farmers near drilling districts, or city dwellers reliant on cheap lamp oil, welcomed policies ensuring consistent supply or fair pricing. By 1900, though no universal blueprint existed, a patchwork of laws, tariffs, and concessions shaped the day-to-day operations of late 19th-century oil producers, reflecting each region's distinct political culture.

7. Intellectual and Cultural Reflections on Oil's Future

7.1. Journalistic and Literary Speculations

Toward the end of the 19th century, newspapers and magazines sometimes featured forward-looking pieces about how oil might shape the coming decades. Some predicted that with more advanced refining, petroleum could replace older fuels more thoroughly. Others warned that known oilfields might run dry, plunging the industry into crises unless new fields were found. There was no consensus; speculation ranged from boundless optimism to cautionary gloom.

Literary works, especially in countries touched by oil booms, occasionally included characters who left everything behind to chase rumored fields. The moral of such stories often hinged on whether the pursuit of oil brought prosperity or ruin, reflecting broader cultural anxieties about modernity's gains and losses.

7.2. Emerging Technical Education

A handful of technical colleges introduced specialized courses covering oil geology, refining engineering, or pipeline management. Younger generations enrolled, seeing the industry as an avenue for innovation and steady employment. This educational shift hinted that oil was becoming a recognized sector worth formal study—no longer a dusty frontier enterprise but a legitimate career path.

Graduates from these programs carried new ideas back to the field, improving drilling efficiency, analyzing reservoir data more precisely, and enhancing refining yields. Although these efforts were small by modern standards, they heralded the intensification of professional expertise that would define 20th-century petroleum engineering and geoscience.

8. Oil and the Edge of the 20th Century: Key Regions Poised for Growth

8.1. North American Outlook

By 1900, North America retained a lead in oil production volume and refining expertise, especially after decades of expansion in the U.S. A cluster of integrated companies, some quite large, had begun to shape national and regional markets. Infrastructure—like pipelines and refineries—had improved, and the public was accustomed to oil lamps and lubricants. Nonetheless, the continent's geological riches were not fully mapped; states like Texas and Oklahoma, soon to be major players, were just beginning to draw serious attention.

The synergy between rising industrial demands, expanded rail networks, and well-organized drilling gave the United States a head start, overshadowing smaller Canadian fields or modest Mexican seepages. All the same, the success story gave impetus to geologists and speculators to keep pushing west and south, eyeing potential new fields that would eventually shift the industry's center of gravity.

8.2. The Russian Empire and Eastern Europe

The Caucasus, anchored by Baku, stood as the second major pole of late 19th-century oil output. Russian industrialization, though uneven, placed a heavy emphasis on harnessing Baku's wells. Railroad expansions linked the region to other parts of the empire, and port facilities on the Caspian Sea allowed seaborne transport. Meanwhile, Eastern Europe's Galicia maintained a steady flow, serving Central European markets with enough volume to stay profitable.

Political tensions within the Austro-Hungarian and Russian Empires sometimes hindered cross-border collaborations. Nonetheless, the success at Baku spurred further geological surveys around the Caspian and beyond. Russian and foreign companies scouted additional basins, hoping to replicate the fortunes made in existing fields. By the century's end, the region's reputation for prolific oil grew, reinforcing a narrative that the Old World could match or rival the New in petroleum wealth.

8.3. Distant Lands and "Sleeping Giants"

Though overshadowed in production terms, other areas glimpsed future significance. In the Middle East—Mesopotamia, Persia, the Arabian Peninsula—scattered drilling efforts, minimal though they were, suggested deeper reserves. Southeast Asia's Dutch East Indies witnessed small but telling successes in Sumatra and other islands. Latin America had likewise seen early tests in locales like Venezuela, albeit with limited follow-through due to political upheavals and transport difficulties.

These "sleeping giants" remained largely untapped, lacking the stable conditions or large capital flows needed for major booms. But the seeds were planted by the late 19th century. Geologists who read about these attempts realized that Earth's petroleum endowment stretched far wider than previously assumed. The stage was set for future expansions in all these regions once technology, finance, and political contexts aligned—well into the 20th century.

9. Summarizing the Late 19th-Century Threshold

9.1. Convergence of Technology, Demand, and Investment

By about 1900, a convergence of factors propelled oil from a secondary resource to one on the cusp of transformative relevance:

- **Technology**: Drilling rigs, pipelines, and refining processes were significantly improved, able to handle deeper wells and yield better products.
- **Demand**: Households worldwide sought lamp oil; factories, trains, and steamships needed lubricants; and experiments hinted at new fuel applications on the horizon.
- **Investment**: Joint-stock companies and cross-border financing fueled broader exploration, integrating local fields into an increasingly global network.

These elements combined to create an industry that, while still behind coal in raw energy output, boasted unmatched versatility and growth potential. The question was no longer whether oil would remain a fixture but how dominant it might become in shaping future economies and societies.

9.2. Legacy of 19th-Century Foundations

All the subsequent leaps—mass automobile production, advanced petrochemicals, global oil politics—rested on the 19th-century foundation. The men and women who labored in rudimentary wells, refined thick crude in open vats, or hammered out pipeline routes laid the essential groundwork. Communities that fought local pollution or improvised minimal safety measures demonstrated the resource's capacity for both harm and benefit.

Though overshadowed by bigger developments in later decades, these experiences shaped local legal precedents, spurred initial environmental awareness, and honed the drilling and refining craft. By the closing moments of the 19th century, the essential puzzle pieces were in place. The next wave—beyond our chronological scope—would see them snapped together into the modern petroleum age, transforming the globe in ways those earliest well-diggers and lamp-oil merchants could scarcely have imagined.

10. Conclusion of Chapter Twenty

Looking back on oil's 19th-century saga, we see a resource that began as a curiosity—extracted by hand near surface seepages, employed primarily for waterproofing or lighting. Over decades, methodical drilling spread, refining techniques advanced, markets for lamp fuel and lubricants expanded, and a global wave of speculation and exploration ensued. By 1900, the sum of these changes had pushed petroleum to the brink of a new epoch.

Politically, states realized oil could enhance economic stability or feed industrial growth. Socially, communities living with oil's benefits and hazards developed complex relationships to the substance, blending hope, caution, and entrepreneurial zeal. Culturally, newspapers and travel writers romanticized or sensationalized discoveries, fueling public fascination with "black gold." Environmentally, local protests and small ordinances signaled that the resource's exploitation carried real consequences for land and water.

All these threads converged as the 19th century ended—establishing the threshold of modernity in the oil domain. Soon, new drilling frontiers would open in Texas, California, the Middle East, and beyond, propelled by more advanced engines, rising urban populations, and far-reaching distribution networks. The 19th-century foundation ensured that when these expansions occurred, the oil industry could scale quickly, forging global corporations and reshaping energy systems worldwide.

This book has traced oil's historical arc up to this threshold, stopping short of the 20th-century transformations. By examining the ancient foundations, medieval continuity, early industrial uses, political jockeying, cultural shifts, and environmental debates, we see that oil's modern role did not appear out of nowhere. It was the outcome of centuries of gradual discovery, scientific trial, small-scale refining experiments, scattered legislative steps, and the cumulative effect of countless individuals seeking to harness a substance that the earth leaked in sticky black pools. In this story, each era contributed a crucial piece—preparing oil, and humanity, for the momentous developments just beyond our chosen historical boundary.

Help Us Share Your Thoughts!

Dear reader,

Thank you for spending your time with this book. We hope it brought you enjoyment and a few new ideas to think about. If there was anything that didn't work for you, or if you have suggestions on how we can improve, please let us know at **kontakt@skriuwer.com**. Your feedback means a lot to us and helps us make our books even better.

If you enjoyed this book, we would be very grateful if you left a review on the site where you purchased it. Your review not only helps other readers find our books, but also encourages us to keep creating more stories and materials that you'll love.

By choosing Skriuwer, you're also supporting **Frisian**—a minority language mainly spoken in the northern Netherlands. Although **Frisian** has a rich history, the number of speakers is shrinking, and it's at risk of dying out. Your purchase helps fund resources to preserve and promote this language, such as educational programs and learning tools. If you'd like to learn more about Frisian or even start learning it yourself, please visit **www.learnfrisian.com**.

Thank you for being part of our community. We look forward to sharing more books with you in the future.

Warm regards,
The Skriuwer Team

www.ingramcontent.com/pod-product-compliance
Lightning Source LLC
Chambersburg PA
CBHW050526240926
51846CB00003B/710
* 9 7 8 3 5 6 5 0 8 4 8 4 5 *